“The Object Lessons series achieves something very close to magic: the books take ordinary—even banal—objects and animate them with a rich history of invention, political struggle, science, and popular mythology. Filled with fascinating details and conveyed in sharp, accessible prose, the books make the everyday world come to life. Be warned: once you’ve read a few of these, you’ll start walking around your house, picking up random objects, and musing aloud: ‘I wonder what the story is behind this thing?’”

Steven Johnson, author of *Where Good Ideas Come From* and *How We Got to Now*

“Object Lessons describes themselves as ‘short, beautiful books,’ and to that, I’ll say, amen. . . . If you read enough Object Lessons books, you’ll fill your head with plenty of trivia to amaze and annoy your friends and loved ones—caution recommended on pontificating on the objects surrounding you. More importantly, though . . . they inspire us to take a second look at parts of the everyday that we’ve taken for granted. These are not so much lessons about the objects themselves, but opportunities for self-reflection and storytelling. They remind us that we are surrounded by a wondrous world, as long as we care to look.”

John Warner, *The Chicago Tribune*

"Besides being beautiful little hand-sized objects themselves, showcasing exceptional writing, the wonder of these books is that they exist at all. . . . Uniformly excellent, engaging, thought-provoking, and informative."

Jennifer Bort Yacovissi, *Washington Independent Review of Books*

". . . edifying and entertaining . . . perfect for slipping in a pocket and pulling out when life is on hold."

Sarah Murdoch, *Toronto Star*

"For my money, Object Lessons is the most consistently interesting nonfiction book series in America."

Megan Volpert, *PopMatters*

"Though short, at roughly 25,000 words apiece, these books are anything but slight."

Marina Benjamin, *New Statesman*

"[W]itty, thought-provoking, and poetic . . . These little books are a page-flipper's dream."

John Timpane, *The Philadelphia Inquirer*

“The joy of the series, of reading *Remote Control*, *Golf Ball*, *Driver's License*, *Drone*, *Silence*, *Glass*, *Refrigerator*, *Hotel*, and *Waste* (more titles are listed as forthcoming) in quick succession, lies in encountering the various turns through which each of their authors has been put by his or her object. As for Benjamin, so for the authors of the series, the object predominates, sits squarely center stage, directs the action. The object decides the genre, the chronology, and the limits of the study. Accordingly, the author has to take her cue from the *thing* she chose or that chose her. The result is a wonderfully uneven series of books, each one a *thing* unto itself.”

Julian Yates, *Los Angeles Review of Books*

“The Object Lessons series has a beautifully simple premise. Each book or essay centers on a specific object. This can be mundane or unexpected, humorous or politically timely. Whatever the subject, these descriptions reveal the rich worlds hidden under the surface of things.”

Christine Ro, *Book Riot*

“. . . a sensibility somewhere between Roland Barthes and Wes Anderson.”

Simon Reynolds, author of *Retromania: Pop Culture's Addiction to Its Own Past*

“My favourite series of short pop culture books”

Zoomer magazine

OBJECTLESSONS

A book series about the hidden lives of ordinary things.

Series Editors:

Ian Bogost and Christopher Schaberg

In association with

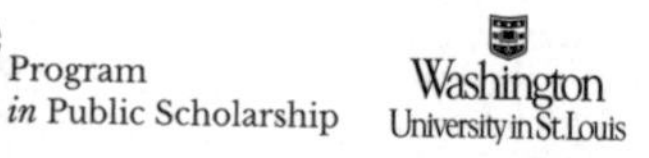

BOOKS IN THE SERIES

Air Conditioning by Hsuan L. Hsu
Alarm by Alice Bennett
Barcode by Jordan Frith
Bicycle by Jonathan Maskit
Bird by Erik Anderson
Blackface by Ayanna Thompson
Blanket by Kara Thompson
Blue Jeans by Carolyn Purnell
Bookshelf by Lydia Pyne
Bread by Scott Cutler Shershow
Bulletproof Vest by Kenneth R. Rosen
Burger by Carol J. Adams
Cell Tower by Steven E. Jones
Cigarette Lighter by Jack Pendarvis
Coffee by Dinah Lenney
Compact Disc by Robert Barry
Doctor by Andrew Bomback
Doll by Maria Teresa Hart
Driver's License by Meredith Castile
Drone by Adam Rothstein
Dust by Michael Marder
Earth by Jeffrey Jerome Cohen and Linda T. Elkins-Tanton
Egg by Nicole Walker
Email by Randy Malamud
Environment by Rolf Halden
Exit by Laura Waddell
Eye Chart by William Germano
Fat by Hanne Blank
Fake by Kati Stevens
Football by Mark Yakich
Gin by Shonna Milliken Humphrey
Glass by John Garrison
Glitter by Nicole Seymour
Golf Ball by Harry Brown
Grave by Allison C. Meier
Hair by Scott Lowe
Hashtag by Elizabeth Losh
High Heel by Summer Brennan
Hood by Alison Kinney
Hotel by Joanna Walsh
Hyphen by Pardis Mahdavi
Jet Lag by Christopher J. Lee
Luggage by Susan Harlan
Magazine by Jeff Jarvis
Magnet by Eva Barbarossa
Mushroom by Sara Rich
Ocean by Steve Mentz
Office by Sheila Liming
OK by Michelle McSweeney
Password by Martin Paul Eve
Pencil by Carol Beggy
Perfume by Megan Volpert
Personal Stereo by Rebecca Tuhus-Dubrow
Phone Booth by Ariana Kelly
Pill by Robert Bennett
Political Sign by Tobias Carroll
Potato by Rebecca Earle
Pregnancy Test by Karen Weingarten
Questionnaire by Evan Kindley
Recipe by Lynn Z. Bloom
Refrigerator by Jonathan Rees
Relic by Ed Simon
Remote Control by Caetlin Benson-Allott
Rust by Jean-Michel Rabaté
Scream by Michael J. Seidlinger
Sewer by Jessica Leigh Hester
Shipping Container by Craig Martin
Shopping Mall by Matthew Newton
Signature by Hunter Dukes
Silence by John Biguenet
Skateboard by Jonathan Russell Clark
Snake by Erica Wright
Sock by Kim Adrian
Souvenir by Rolf Potts
Spacecraft by Timothy Morton
Space Rover by Stewart Lawrence Sinclair
Sticker by Henry Hoke
Stroller by Amanda Parrish Morgan
Swimming Pool by Piotr Florczyk
Traffic by Paul Josephson
Tree by Matthew Battles
Trench Coat by Jane Tynan
Tumor by Anna Leahy
TV by Susan Bordo
Veil by Rafia Zakaria
Waste by Brian Thill
Whale Song by Margret Grebowicz
Wine by Meg Bernhard
Concrete by Stephen Parnell (Forthcoming)
Fist by nelle mills (Forthcoming)
Fog by Stephen Sparks (Forthcoming)
Mask by Sharrona Pear (Forthcoming)
Newspaper by Maggie Messitt (Forthcoming)
X-ray by Nicole Lobdell (Forthcoming)

Space Rover

STEWART LAWRENCE
SINCLAIR

BLOOMSBURY ACADEMIC
NEW YORK • LONDON • OXFORD • NEW DELHI • SYDNEY

BLOOMSBURY ACADEMIC
Bloomsbury Publishing Inc
1385 Broadway, New York, NY 10018, USA
50 Bedford Square, London, WC1B 3DP, UK
29 Earlsfort Terrace, Dublin 2, Ireland

BLOOMSBURY, BLOOMSBURY ACADEMIC and the Diana logo are trademarks of Bloomsbury Publishing Plc

First published in the United States of America 2024

Cover design: Alice Marwick

A catalog record for this book is available from the Library of Congress

ISBN: PB: 978-1-5013-9995-4
ePDF: 978-1-5013-9997-8
eBook: 978-1-5013-9996-1

Series: Object Lessons

Typeset by Deanta Global Publishing Services, Chennai, India
Printed and bound in Great Britain

To find out more about our authors and books visit www.bloomsbury.com and sign up for our newsletters.

For Guy and Irene Cooper

Who taught me the beauty of science

For my father

Who taught me to drive, and to persevere

CONTENTS

PART I

What do you build with sails for flight?

I build a boat for Sorrow:
O swift on the seas all day and night
Saileth the rover Sorrow,
All day and night.

—William Butler Yeats, "The Cloak, the Boat and the Shoes"

1 MOONSCAPES

Alan's farmhouse looked as rustic as ever. If I hadn't known better, the lingering smoke in the air would've triggered memories of barbecues, bonfires—not existential catastrophe. It wasn't until my stepfather Loy and I reached the crest of the drive that the aroma aligned with our surroundings, and the reptilian green of the orchard's leaves morphed to charred browns and blacks.

The Thomas Fire left its mark on the Fogliadini ranch. The bark of a nearby palm tree looked like blackened leather. A horseshoe nailed to it for luck was oxidized orange. At the end of the drive, a decimated tractor's seat and tires had melted away—its rusty frame slumped like a heap of bones.

In December 2017, the Santa Ana winds raged southwest from the Mojave Desert to the coast. On the east side of Ventura County, those winds forced two powerlines into contact, depositing molten material onto what fire investigators call a "receptive fuel bed." Within hours, the fire burned across twelve miles, merging with a second fire north of the mountain town of Ojai.

I was out of the country when I heard the news, and I called my mother to see if I should fly home. She told me that my grandmother, my uncle Alan, his wife, my sister, her fiancé, his parents, a friend who'd been partying with them when the fire broke out, five of my nieces and nephews, three dogs and two cats were already staying with her.

Besides, she reminded me, "There's nothing you can do."

Meanwhile, the pictures filling my Facebook feed were otherworldly. Palm trees burst into flames, pyrocumulus clouds mushrooming above the foothills. Firefighters have a word for such total devastation: *moonscaping.*

Ventura burned for thirty-nine days before I finally caught a flight home. I was relieved to see my family, and to know that, for them at least, things had turned out ok. By the time I touched down, the fire destroyed more than 1,000 structures and forced 100,000 people to evacuate. Two people, a firefighter and an elderly woman fleeing in her car, were killed. For a brief moment, the Thomas Fire was the largest wildfire in the state's history.

Then came the rains. A deluge of floodwater sent a river of boulders and debris through Montecito, damaging or destroying hundreds of homes and leaving twenty people dead in the mud. The 101 Freeway was blocked for over a week.

Uncle Alan lived north of Ventura, just beyond where the mudslide had severed the freeway. He'd almost lost his ranch in the fire, so I'd been anxious to pay him a visit and see the orchard and the old house where I'd spent so much of my childhood.

When I stepped out of Loy's truck, the tranquility of the ranch blanketed me. A light wind rustled the trees and reminded me of how this place, like so many rural enclaves, possessed a timeless element. Years pass in endless cycles of irrigating and picking, and to spend even some of your days on that land is to occupy a small Eden.

Alan wasn't there when we arrived, but soon his truck rolled into view. He stepped out of the cab and Loy shook his hand.

"Brought you someone."

"You're a man now!" Alan hugged me, then we looked out over the orchard. A sliver of ocean shimmered down the valley, and the sun haloed the trees in golden-hour light as it sank beneath the horizon.

Loy and Alan bullshitted as I examined a nearby tree and plucked its shriveled leather fruit, halving it with my pocketknife. Its creamy yellow-green flesh radiated from the pit like a horticultural corona. When I tossed it to the floor Alan's terrier dug his nose into the meat.

"The dogs love 'em," Alan said.

The fire reached the ranch about four days after it broke out, burning the upper end of the grove and taking out a power line. Smoke saturated the walls of Alan's home, but his main losses were in his barn. From the outside, it looked untouched, but when Alan opened the door a cloud of ash blossomed. The barn's corrugated aluminum acted like a convection oven, tempting the flames toward the crack beneath the barndoor. Alan pointed toward the soot-path tracing a clockwise vortex from floor to ceiling. His finger tracked the corkscrewed trail of charred photographs, tools, a motorcycle frame, and some wire art that Alan's father made. The model planes hanging from the ceiling looked like they'd taken flack.

"This was an ass-whoopin'," Alan said. "Take a look at this." Next to the door sat a cardboard box of photographs. Alan lifted it onto his workbench. "This sat in that corner the whole time. Never knew until we discovered it in the clean-up."

In one picture, a young version of my father pitched a horseshoe with one hand and gripped a beer with the other. In another, Alan and his half-brothers, my uncle Dickie and my grandpa Sandy, leaned over the hood of a '56 Chevy gasser with the names *Sinclair*, *Nelson*, and *Fogliadini* painted on the side. Rifling through the box, we ventured deeper into our past. Pictures belonging to Alden Fogliadini, Alan's father—my step-great-grandfather. We called him grandpa Fogi.

The first picture was a massive platform that rolled along four sets of railroad tracks. Alan explained it was the crawler-transporter (apparently the largest self-powered land vehicle in the world upon completion in 1965). The platform conveyed every NASA spacecraft from the Saturn IV to the space shuttle to the launch tower from the Vehicle Assembly Building.

In the next picture a row of men in lab coats stood behind what looked like a go-kart, but was actually one of the original prototypes of the lunar roving vehicle, or LRV, more commonly known as the moon rover. Standing on the right with his familiar spectacles hanging off his nose and a disheveled donut of hair around his bald crown was Fogi, a machinist with a high school education, who didn't work directly for NASA but for General Motors. As we flipped through more pictures of the rover intermingled with Polaroids from family cookouts, I considered how such a singular object had been borne of the labor of people from such varied walks of life. I remembered the last time I saw Fogi, when I was sixteen and visiting the ranch with my father. He didn't say anything about the moon or the rover. He was just watching the game and telling my father about how important it was that I play football. With his country drawl and wild hair poking out beneath the brim of his Stetson, he scared me a little.

Had the Thomas Fire consumed Alan's barn, we would never know those pictures existed. That a box of chemically saturated paper survived a burning barn almost suggested divine intervention—or the sheer luck that the flaming vortex had risen clockwise in the barn in deference to the Coriolis effect, preserving that little nook near the door. Whatever your conviction, their survival betrays the inexorability of entropy. This became increasingly apparent to me as I wandered Ventura in the aftermath of the fire, marveling at how much could be incinerated so quickly.

2 AUTHOR'S CONFESSION

This book's title is a misnomer. There are no "space rovers." When a rover is traveling between Earth and another celestial body, it's probably more appropriate to think of it as a spacecraft—or cargo. Even upon landing, a rover is still a specialized spacecraft. The preferred term of art is "planetary rover." But thinking of these objects momentarily as "space rovers" helps distinguish them from their more terrestrial variants: Land Rovers or Range Rovers, your neighbor's dog Rover, or "red rover" the children's game (likely derived from the pirate in James Fennimore Cooper's novel of the same name).

In the case of this book, planetary rovers—space rovers—are machines primarily built to explore "planetary celestial mass bodies," such as our own Moon, Mars, or, in at least one case, large objects in the asteroid belt. But more broadly, we might think of rovers as vehicles designed for places that are inhospitable to humans. Whether manned or autonomous, the term "rover" tends to denote a vehicle that either protects its occupants from the surrounding environment, aids in exploration, or takes a human's place in a hostile or inaccessible environment.

To date, at least ten rovers have explored the Moon and Mars, with the latest addition being Chandrayaan-3, ushering India into the small realm of planet-roving nations (Russia, the United States and China) in July of 2023, and taking the mantle as the first rover to reach the Moon's south pole, a site coveted for its deposits of frozen water. Still more rovers have been designed for other planets, moons, and asteroids, but as yet (though this will likely change soon) none have been successfully deployed. Some were cancelled, others crashed; space is hard.

Rovers have also been used here on Earth, as training vehicles for Apollo astronauts and Mars missions, for Antarctic expeditions, and even to help resolve some of the worst man-made catastrophes—bulldozing through nuclear waste or probing earthquake wreckage and collapsed mines. Boston Robotics has even gone viral with bi-pedal and quadrupedal rovers. A troupe of their canine-inspired "Spot" rovers appeared on *The Tonight Show with Jimmy Fallon*, performing a simultaneously hypnotic and disturbing choreographed dance to BTS's "IONIQ: I'm on It."

Often, technology developed for one of these purposes is adapted for another, and an ecology has emerged of evolving objects across which we can assign the categorical designation: *Rover.*

As far as space hardware goes, rovers are unique. I'd argue that no rocket, spaceship, lander, satellite, probe, or space station was ever quite so anthropomorphized as some of the more well-known rovers. When the Mars Curiosity rover first touched down, its Twitter account tweeted, "I'm safely on Mars. GALE CRATER I AM IN YOU!!!#MSL." The account has four million followers. One of my favorite Curiosity tweets: "Here I go again on my own. Down the only road I've ever known. Out of safe mode—back to work." Curiosity's successor, Perseverance, currently has three million followers. And recently tweeted, "It seems like only 'yestersol' since I last saw my original landing site."

What is it about rovers that makes them such singular objects of our affection? In some cases, it's by design. The dual cameras on Opportunity and Spirit are constructed intentionally to look like a pair of eyes. The first-person tweets deliberately garner empathy. The vehicles' squat six-wheeled bodies with their flat hexagonal arrays of solar panels on their "backs," their long "necks," and tail-like antenna suggest some kind of robo-pet, with names that sound as if they were bestowed by Mr. Rogers himself. (In that same vein, a public poll conducted by the China National Space Administration (CSNA) to name their rovers produced the name Yutu, the pet rabbit of Chang'e, the goddess of the Moon in Chinese mythology, and the name of the landers which would transport Yutu and Yutu-2 to the Moon in 2013 and 2018.) But beyond all that, there may be something about the purpose of these rovers that innately appeals to the human condition—whatever that condition may be.

A rover, by one Oxford definition, is a person who spends their time wandering. To wander is to move aimlessly. All space rovers have possessed clearly defined objectives, but it is the uncertainty of their missions that make them so compelling. It's been said that when searching for intelligent life, you should look for straight lines, geometric regularity. There's probably truth to that. But when looking for signs of intelligent robots (or robots operating off of the inputs of an intelligent being), maybe it's best to look for erratic trails. The paths charted by lunar and Martian rovers aren't straight lines; they're meandering paths, diversions toward objects of interest.

Humans first realized there was something special about a handful of stars when they identified their unpredictable movements across the night sky, rising, falling, and sometimes swooping in retrograde motion in relation to the other "fixed" points of light. The Greeks named these points *planetes*, meaning wanderers. Upon those planets they projected their gods.

The word "rover" comes from the Middle Dutch *roven*, "to rob," and the Danish and Norwegian *røver*, "robber, thief, highwayman, brigand." As I've mentioned, a rover is also a pirate. When astronaut Mark Watney departs the HAB to commandeer the Ares 4 lander in the film *The Martian,* he points out that international law forbids countries from claiming territory off-Earth, which makes Mars "international waters" governed by maritime law. So, when Watney boards Ares 4 without explicit permission, he's technically pilfering a vessel in international waters, "which, by definition, makes me a pirate."

Matt Damon is Mark Watney is Space Pirate is Space Rover.

Through these definitions, a long-sustained tension emerges associating aimless wandering with vagrancy, even criminality. The moment one moves without purpose or permission, the movement becomes suspect.

This may be part of what compelled the United States and Russia to willingly commit near-limitless resources to landing a human on the Moon, but act hesitantly toward further investment in space exploration at the same levels after that objective had been achieved.

Maybe there is something peculiar, in a profound evolutionary sense, about our species' urge to wander. In her book *The Sixth Extinction: An Unnatural History*, Elizabeth Kolbert explores the concept of "The Madness Gene," which contemplates why archaic humans like *Homo Erectus* or Neanderthals never crossed oceans to discover unknown shores. "It's only fully modern humans who start this thing of venturing out on the ocean where you don't see land," Kolbert writes. She continues,

> Part of that is technology, of course; you have to have ships to do it. But there is also [. . .] some madness there. You know? How many people must have sailed out and vanished on the Pacific before you found Easter Island? I mean, it's ridiculous. And why do that? Is it for glory? For

> immortality? For curiosity? And now we go to Mars. We never stop.

Donald Goldsmith and Martin Rees counter in *The End of Astronauts: Why Robots are the Future of Exploration* that it's easy to assign a teleological imperative to exploration without any supporting evidence:

> The desire to explore is not in our destiny, nor in our DNA, nor innate in human cultures. The first assertion has only mysticism in support, the second has no genetic evidence, and the third encounters negative evidence around the world. If sending humans to Mars were "our destiny," we would have no reason to hurry down this path, as many insist we must, since we would be sure to get there eventually. If our DNA someday revealed a genetic bias toward exploration, this could have resulted from natural selection in the descendants of those who had engaged in exploration and survived. And while some cultures have proven enthusiastic adventurers and explorers, many others have not. For example, Polynesians of past centuries were fearless explorers, while their Chinese contemporaries found satisfaction remaining within the Middle Kingdom.

Maybe the "need" to explore—or rather the "need" to conquer or colonize—is a survival mechanism adapted among some cultures and not others, or simply another ancillary consequence of our political institutions. Or maybe we *are* just evolutionary pirates, venturing across interplanetary waters to plunder our way into a broader ecological niche, consuming everything in our path along the way, voyeuristically cataloguing the unknown to satiate our own curiosity regardless of the consequences. The thought calls to mind Susan Sontag's claim in *Regarding the Pain of Others* that "The photographer was a rover, with wars of unusual interest (for there were many wars) a favorite destination." And what is a planetary rover at its core if not a photographer navigating a literal no-man's land.

Or maybe to be a rover is just an expression of our fundamental right to movement as inscribed in article 13 of the Universal Declaration of Human Rights.

No matter the vantage point, when we look at the erratic movements of rovers across the surfaces of other worlds, can we help but connect them to our own migratory wanderings across the surface of this—our own—planet? In both cases, our first steps, or tracks, across previously unencountered terrains, always pave the way for others to follow—and with that comes the potential to bring the best and the worst of what our wanderings have delivered to every distant shore upon which we've ever landed.

3 MOONBEAMS

Late nights in the backseat of my mother's Chevy Astro Van, I'd press my forehead against the glass, watching the condensation swell and contract with my breath. In Ventura, long country roads cut straight lines through dark strawberry fields and lemon groves. Beyond the streetlights and telephone poles, I'd watch the Moon drifting imperceptibly across the sky above the mountains.

I've always been a Moon gazer. It surprises me that some people aren't. Its phases dictate the rhythms of our calendar. Its orbit and mass compels the oceans' rise and fall. Its full disk in June coaxes horseshoe crabs to breed in the moonlight and courts countless young human lovers into each other's embraces.

The Moon is a shy child's companion. I was the only boy in a family of five. Our parents divorced when I was four, and we lived with our mother, who worked multiple jobs to support us. We'd see Dad on the weekends, but he was a welder who often worked off-shore or in the oil fields. In that loneliness, I often looked to the Moon, especially in those early years when my sisters and I all slept in one bed and

I'd fall asleep to the rumble of the train rolling through our backyard, and the sight of the Moon out the window.

I always thought the man in the Moon possessed a pained expression. In Jewish lore, he's the man God sentenced to death by stoning after he was caught gathering sticks on the Sabbath. Romans considered him a sheep-thief. Some medieval Christians claimed he was Cain, the Wanderer, forever doomed to circle the Earth. Carol Ann Duffy portrays a sorrowful matriarch lamenting a planet in crisis in "The Woman in the Moon":

> . . . When your night comes,
> I see you staring back as though you can hear my
> Darlings,
> What have you done, what you have done to the earth?

All of this makes the Moon a fine companion for a lonely child. She doesn't placate or humor you; she commiserates, weeps, ebbs and flows with your joy and lament.

I also felt a special connection, spanning that two-hundred-and-fifty-thousand-mile void, with those three vehicles abandoned on its surface. I'd strain my eyes to see them, the distance diminished and the objects rendered immediate by the knowledge that something Fogi placed his hands on was resting on its surface.

I remember an elementary school picture day around second grade. We lined up against the cafeteria wall and Miss Martinez went down the line encouraging us to smile.

"Let me see those teeth!"

I shook my head.

"C'mon! You've got to be ready for the camera!"

"I *am* ready. I just don't wanna smile."

"It'll make your mom happy. She wouldn't want you frowning, would she?"

"We don't have to buy them."

Miss Martinez walked on. One by one, as each student took their turn, I braced for the photographer's charm offensive. I took my place and stared stoically into the camera. The photographer offered me a Tootsie Roll.

"I don't like chocolate."

Miss Martinez whispered into his ear. He grinned.

"Moon rover," he said.

Against my will, I smirked, and he snapped the picture.

I'm not sure how Miss Martinez knew those words would get to me. I must've betrayed my enthusiasm in a moment of unguarded excitement. The smile was sincere but involuntary, and that moment stayed with me. Whenever it comes back, I remember that floating lunar dreamscape of my youth. No other satellite in our solar system looms so large in its planet's sky. If the Moon were smaller, farther, less solemn in its disposition, humans might never have felt tempted to reach for it. Maybe the constant allure of that lunar frontier compelled so many to cross vast oceans in search of the unknown.

Because the Moon is there, eternally there, it insists on something beyond the beyond.

4 SPLENDID TERROR

Forget the Moon you know: that barren sphere pock-marked with craters, its whisp of an atmosphere comprised of solitary atoms and dust motes. Envision the turn-of-the-century Moon—a mystery whose formation, composition, and habitability remained debatable.

Into that Zeitgeist, Polish author Jerszy Żuławski offered one of the first and strangest lunar rovers in his 1903 novel *On the Silver Globe*, in which an international crew of adventurers embarks on a surreal odyssey to the Moon in search of utopia. Across the lunar terminus, in their hearts and minds, they smuggled a fantasy:

> Perhaps—they sometimes dreamed, drunk with the greatness of their enterprise, just perhaps—on that mysterious side of the Moon, they would find a magical and strange paradise. Perhaps, it would be different from the Earthly one, but hospitable? Then, they were dreaming about calling other volunteers eager to travel through hundreds of thousands of kilometers to the Moon. They thought about starting a new society there, a new humankind . . . perhaps happier . . . perhaps . . .

Żuławski's novel sounds like a Vernesian adventure pursued with unquestioning faith in technology. Its crew even has the whimsical composition of a Wes Anderson film: the Irish astronomer O'Tamor; the love-struck English physician Thomas Woodbell and his beautiful (and apparently pregnant) Indian fiancée Marta; the Brazilian-Portuguese engineer Piotor Varadol; and of course, their dog Selena. Their ship, a hermetically sealed vessel that converts from spacecraft to rover, is stocked with a year's provisions and a respectable library of natural and human history, poetry, and the Bible. The electric vehicle is outfitted with steel wheels, tractor treads and grappling claws, and pays homage to Vernes's bullet-shaped projectile in *From the Earth to the Moon* while anticipating the rovers that would populate sci-fi and popular science in the coming decades, such as Hugo Gernsback's spherical steel chamber floating from Earth to the Moon in *Baron von Munchausen's Scientific Adventures* (1915); or Homer Elon Flint's Seussian "walking rover" in his 1923 novel *Out of the Moon*.

These rovers function as covered wagons pioneering the celestial frontier, capitalizing on the Moon's untapped resources and helping establish the first lunar outposts. But such purposes are subverted in Żuławski's novel. After the vessel crashes, killing O'Tamor and injuring Woodbell, the remaining crew embark on a brutal pilgrimage across Mare Imbrium in their proto-rover, hoping that the far side's suspected atmosphere will spare them from death.

The near side of Żuławski's Moon reflects much of the science of his time. The expedition navigates a gravity field one-sixth the Earth's, the difficulty of ascertaining distance or scale in a monotonous landscape, the threat of the vacuum, and the extreme temperature shifts between sunlight and shadow. While Korecki's journal describes the heavens as if "looking at the sky through a wonderful stereoscope," the moonscape compels him to invoke Dante: "Hell couldn't be more frightening than the one I was looking at now. The smoke of the volcanoes looked like rows of damned spirits twirling around the horrible shape of Lucifer taken upon by one of the volcano cones."

The Moon only further devolves into dystopia when the crew intercepts a telegraph from a second ill-fated Moon-bound craft. A streak shoots across the sky as the vehicle crashes in the distance. The thought of recovering the second vehicle's oxygen overshadows the dismal sight of the mangled corpses of the Remonger brothers, the crew manning the craft. Korecki and the others bury the bodies, scavenge what they can and continue their journey.

As their vehicle traverses the mare, they discover formations that eerily resemble a ruined city. Woodbell leaves the vehicle to investigate, but as he nears the formations, he freezes. Two shadows appear. "No, not shadows!" he recounts later, "Two people or two corpses or maybe ghosts stepped out from under the gate and were walking straight towards me." The Remonger brothers, holding hands, their bodies bloody and swollen, telling Woodbell when and how he will die, and that "there was no leaving Earth without punishment." Woodbell runs back toward the vehicle, tripping over a rock and cracking his helmet as he collapses in the regolith.

This moment will forever haunt Korecki. His thoughts will continuously return to the folly of their pursuit—the dangers of that blind faith in science and the delusion of utopia. The crew's vehicle begins to function like Charon's boat ferrying its doomed passengers between the "real" world of the lunar near side and the allegorical dreamscape of the far side, where they will discover water lakes, petroleum pools, edible mosses and small hard-shelled creatures that at first augur hope, but ultimately intimate an inescapable purgatory.

Woodbell won't survive to see the hidden hemisphere, but Martha will bear his twins, who in turn will interbreed generations of diminutive, short-lived offspring. Korecki, eventually the sole surviving Earthling, will try to preserve the connection the "Selenites" have to their ancestral planet by reading from his books, but each rapidly succeeding generation becomes further alienated. Earth, hidden below the horizon of the tidally locked Moon, becomes mythical, and Korecki, the "Old Man," becomes its god. The Selenites internalize a Messianic prophecy of an Earthling coming to bring them home.

Żulawski's arguments portend contemporary criticisms of space exploration; questions of whether our exploits in space come at the expense of life on Earth; of the quality of existence in a space colony; of who gets to be our interplanetary voyagers; of what we export when we arrive on another planet and what gets left behind when we leave Earth; and at what point we cease being Earthlings, or even humans, at all.

In a 2022 reflection on his grandfather's work, Adam Żuławski observed how *On the Silver Globe* blended "high-adventure romance stories with the anxiety of the beckoning age of modernism," forcing idealistic characters to confront their naive delusions:

> Stretched between melancholia after an evanescing sense of wonder and a hope for the possibility of resetting of human history without erasing cultural memory, Żuławski's adventurers quest after the "grail" of re-enchantment. Their pursuit takes them across landscapes that keep reminding them—like the spectral sphere of the Earth haunting Korecki's expedition—that even when we manage to escape to other worlds and to the future, we will always find that our history got there before us.

From a mountainous village in the occupied territory that would one day be Poland, Żuławski witnessed the birth of a Soviet Union that anticipated the end of history and the arrival of socialist utopia. That utopia was intertwined in some ways with Russian Cosmism, a peculiar movement prophesying universal human resurrection through technological and scientific progress. Originating in the early nineteenth century, Cosmism profoundly influenced the Soviet space program—particularly through its primary innovator and evangelist, Konstantin Tsiolkovsky, one of the fathers of rocket science—at the turn of the twentieth.

Tsiolkovsky conceived of rockets with steering thrusters, multi-stage boosters, space stations, airlocks, and even, in 1918, a Moon rover. But his greatest contribution, published in the same year as *On the Silver Globe*, was his rocket equation, defined by three immutable variables: the energy expenditure against gravity, the energy available in your rocket propellant, and the propellant mass fraction (how much fuel is needed compared to the mass of the rocket). So immutable are these constraints that they are referred to as "the tyranny of the rocket equation," though Tsiolkovsky might likely have thought about them as liberating—a foundational constraint that, if mastered, would open the universe. Tsiolkovsky combined his scientific reasoning with Cosmism, offering space travel as the mechanism through which humanity would extinguish suffering, ushering in a transhuman future as we colonized the Milky Way.

"Earth is the cradle of humanity," he wrote, "but one cannot remain in the cradle forever."

These ideas were and are not limited to eastern Europe. The thread runs through the Star Child of Arthur C. Clarke and Stanley Kubrick's *2001: A Space Odyssey*, in the Afrofuturist divinations of Sun Ra's *Space is the Place,* in conversations about artificial intelligence or the singularity, or whenever certain futurist billionaires suggests that humanity's survival depends on becoming a multi-planetary species.

But to Żuławski, such astral utopianism is just repackaged earthbound ideology: exporting the rapture to the Moon. (Slavoj Žižek describes Cosmism as "a strange combination of vulgar materialism, agnostic spirituality, which formed an occult shadow ideology, the obscene secret teaching of soviet Marxism.") On Earth or the Moon, situating paradise on the edge of civilization, imminent but never now, merely avoids actually confronting society's problems.

Żuławski saturates Korecki's journal with epiphanies about the folly of such thinking: "I repeat to myself a hundred times that I will be dying with these people as a voluntary sacrifice of that all-powerful desire for knowledge. A desire that tore us all away from Earth and dropped us on this inhospitable globe." To be severed from Earth, he ultimately realizes, is to be severed from one's humanity. Korecki will live a long, isolated existence, but the Selenites will live nasty, brutish and short lives anticipating a terrestrial messiah's deliverance from an infernal Moon to an Earthly Eden.

Apollo astronauts have often said that in going to the Moon, they discovered the Earth, comprehending in a new way its beauty and fragility through what's become known as the overview effect. Eugene Cernan remarked of the experience of being in space, orbiting the Earth, that "You have to, literally, just pinch yourself and ask yourself the question, silently: Do you really know where you are at this point in time and space, and in reality and existence? When you look out the window and you're looking back at the most beautiful star in the heavens—the most beautiful because it's the one we understand and we know—it's home, humanity, people, family, love, life. And besides that, it is beautiful." It's this deep revelation about how precious this place called Earth is. *On the Silver Globe* feels like an early examination of that epiphany (or perhaps that horrific realization) in the minds of individuals who know that they will never return home to appreciate that fragile beauty, or to engage the world with that newfound sense of urgency. Korecki set out believing that their vehicle was like an ark, or perhaps a seed—some pan-spermatozoic cell that would reap a new humanity on the Moon—only to realize that to be human is to be an Earthling, and that seed might not, or perhaps should not, germinate in another world's soil.

5 CALIFORNIA STARS

My father has a sepia-tone picture of a family, the Odells, in front of an old car from the 1920s or '30s. No one's smiling. Everyone looks sand blasted. A little girl, Lorraine Odell, squints in the sun. The Odell's were from Shawnee, Oklahoma, and had either just arrived or were just departing for California, a few among the 2.6 million migrants escaping the Dust Bowl and the Great Depression, fleeing the Midwest. Their cars became their mobile homes, life savings, refuges. Woody Guthrie called them ramblers. I call them rovers. When they arrived at the California border, they were met with signs that said *No jobs in California. If you are looking for work—KEEP OUT.* Many were sent to work camps. Many others found nothing there for them.

Lorraine and her family survived the lean years and settled down in Carpinteria, where she would one day marry a man named Alden Fogliadini, the son of Italian immigrants. Together, they bore two children and adopted another: Alan, Dickie, and Stewart. Those post-war years brought freeways, the military industrial complex, and the nascent space industry to southern California. The boys fell into the burgeoning southern California car culture, building hot rods and racing them down Carpinteria's main drag, and their parents built careers at General Motors. Gigi was a secretary, Fogi a machinist. Together, they scrimped and saved for a few acres of land in the Carpinteria hills. They planted avocadoes, grew old, and helped put a car on the Moon.

6 AUTOPIA

In 1965, Eugene Cernan and half a dozen other Gemini astronauts sat across from Dr. Wernher von Braun, first director of the Marshall Space Flight Center, for an informal meeting.

"Don't you worry about getting to the Moon," von Braun said, "I will get you there. It's what you do when you get there that's important." After a pause, von Braun added, "You will probably be driving a car on the Moon." Cernan recalled the audacity of the proposition. "Going to the Moon was still a long way off, and he was talking about driving cars a quarter of a million miles away."

Von Braun had been popularizing his vision for more than a decade. Through the 1952 *Collier's* series "Man Will Conquer Space," he emerged from his military cloister to publicize a vision of human space exploration. In this vision, rovers would be critical:

> The first equipment brought out of the cargo ship is one of three surface vehicles, tanklike cars equipped with caterpillar treads for mobility over the Moon's rough

> surface. The pressurized cylindrical cabins hold seven men, two-way radio equipment, radar for measuring distances and depths, and a twelve-hour supply of oxygen, food, water and fuel. [. . .] As the moon car has been set down and checked, a search party boards it to scout out a suitable crevice for the camp site. They drive off in a spray of dust which settles almost immediately, like the bow wave of a motorboat.

At the time of publication, von Braun was still blowing up test rockets at the Army Ballistic Missile Agency (ABMA) in Huntsville, Alabama. No matter. The Moon, he insisted, was only ten years away. That confidence, combined with the clarity with which he communicated his vision, underscored why *Collier's* editor Cornelius Ryan considered von Braun "one of the best salesmen of the twentieth century," deserving a place in the pantheon of influential showmen from P.T. Barnum to Walt Disney. They were not so much the entertainers, but the architects of spectacle, defining their industries through the cultivation of a meticulously curated image.

Von Braun was uniquely positioned to both articulate America's future in space and dislodge it from fantasy, and it happened that around the time that von Braun was promoting space exploration in *Collier's*, Walt Disney was introducing "the happiest place on Earth" to the American public. In 1954, ABC launched *The Wonderful World of Disney*, a weekly series highlighting each of Disneyland's four realms: Fantasyland, Adventureland, Frontierland, and Tomorrowland. Walt had been struggling with how to present Tomorrowland, whose premise was grounded perpetually in the future; "the only problem with anything of tomorrow," he later said, "is that at the pace we're going right now, tomorrow would catch up with us before we got it built."

In 1955, space seemed like the most reliable "tomorrow" Disney could bank on, and von Braun's optimistic prognostications aligned with Walt's concept. *Man in Space* aired in 1955, with each of the three space installments (the second and third being *Journey to the Moon* and *Journey to Mars*) furthering the public's frame of reference for space exploration, while also molding von Braun's persona. Referred to only as a "German rocket scientist," the series elides his past as an SS officer whose V-2 missiles killed thousands of civilians and prisoners during their production and use in the Blitzkrieg. Von Braun had maintained since his extradition that he'd been an apolitical scientist whose career under a totalitarian regime depended on allegiance, and that he'd intended his research for peaceful pursuits—no matter how the Nazis used it. This interest so angered the Gestapo that they imprisoned him for two weeks after he allegedly drunkenly presumed Germany's defeat at a party, confessing that he'd rather be building a spaceship.

Questions of von Braun's complicity would remain forever, but they didn't outweigh his strategic value. So, the US expunged von Braun and his colleagues' Nazi records during their extradition under Operation Paperclip. Upon arrival in the US, von Braun cloaked himself in a veil of American patriotism, and his collusion with a murderous regime was interred in secret files until the 1980s.

Man in Space was a prescient representation of the architecture of the space age, but it also betrayed an American desire, or perhaps a desire firmly entrenched in American media, for untroubled narratives and historical myopia. The Disney version of space is a demilitarized realm of high adventure, appealing to aspirations rather than fears. In Disneyland, von Braun and his V-2 can arrive in our living rooms from the world of tomorrow unburdened by the past.

The radar bleeps from a grapefruit-sized satellite named Sputnik two years after *Man in Space* aired peeled back that Disney veneer. Suddenly, the fear of orbiting nuclear weapons amplified anxieties surrounding the nuclear stockpiles amassing between two ideologically opposed super-states. Within a year of Sputnik, the US military commissioned a study on the scientific and strategic merits of nuking the Moon, a project that would remain unknown to the public until 2000, and in which a young Carl Sagan was involved in predicting the effects of such a blast.

Come 1959, the ABMA commissioned Project Horizon, a study examining the feasibility of a manned military base on the Moon within the decade. This study included the US's first serious analysis of a lunar transportation system, assigned to the Army's Transportation Corps, who in turn sought advice from General Motors and the head of their Special Vehicle Design Unit, Sam Romano. Romano would recruit Mycislav

"Greg" Bekker, a professor at the University of Michigan who specialized in offroad mobility and soil mechanics. Bekker hailed from a small village in southeast Poland, but fled the country after Hitler's invasion. As Panzer tanks rolled over Europe, he pursued a deeper understanding of the relationship between vehicles and the terrain, seeking a set of underlying principles that would help the field transition from trial and error to hard science—offering an advantage in engineering effective military equipment for the muddy battlefields of Europe.

Working alongside Bekker was Ferenc "Frank" Pavlics, a Hungarian engineer who fled Budapest with his wife after the uprising of 1956. The couple braved a muddy, mine-laden no-man's land patrolled by troops and attack dogs to sneak across the border, boarding a Navy troop ship weeks later and spending nine days crossing the north Atlantic before arriving at Camp Kilmer in New Jersey, where Bekker recruited Frank for his Detroit lab.

The Army Transportation Corps envisioned Project Horizon's "Moon Truck" as a large, roughly two-thousand-pound metal pressurized cabin on caterpillar treads that could ferry soldiers and equipment, powered by rechargeable batteries, with a fifty to seventy-five mile range. Bekker was eager to apply terramechanics to an off-world vehicle's construction. His research, laid out in his foundational 1989 book, *Theory of Land Locomotion*, looked to how animals evolved from amphibians squirming through mud on their bellies into ambulatory creatures who walked, ran, and jumped, proposing that "in general, it must be said that the strength and deformation characteristics of various soils have played a definite role in the [. . .] form, size and weight of the animal body."

The problem with building a vehicle for the Moon was that no one really understood its composition. Many scientists suspected that the surface was covered in a loose dust-like material, with some suggesting astronauts and spacecrafts might sink into the regolith, though that theory was largely dismissed. Bekker and Pavlics tested various hypotheses in soil bins using mixtures of wheat flour and pumice to approximate theoretical lunar soils and their interactions with caterpillar tracks, multi-wheeled articulated vehicles and even, at one point, an Archimedean screw.

It was a strange exercise, drawing a throughline between Bekker's war-time quest for principles of soil mechanics back to that novel he'd encountered as a child, about an experimental vehicle navigating across the silver globe. It must have felt surreal to be professionally engaged in what was once science fiction.

Although Project Horizon never extended beyond that feasibility study, Romano, Pavlics and Bekker would continue their mobility research at GM's new Defense Research Laboratories in Santa Barbara in 1961, developing concepts for an experimental unmanned rover for NASA's Surveyor program, including one six-wheeled articulated rover that could climb objects more than twice its height (though the US would never use a remote rover on the Moon).

Within the next few years, the US Ranger, Lunar Orbiter, and Surveyor missions (and the Soviet Luna missions) brought the moonscape into clearer focus. Ranger probes barreled into the surface, camera shutters firing rapidly until just milliseconds before impact; the Lunar Orbiter provided satellite reconnaissance; and by the mid-60s, Surveyor 1 and Luna 9 made the first soft-landings on the Moon, confirming a land covered with a layer of fine yet sturdy dust, stable enough to support a landing craft.

These findings kept von Braun's dream of a "moon truck," alive, and throughout the early 60s he maintained a mission profile utilizing two Saturn V rockets for each moonshot: one to deliver the crew and another to haul a truck that housed their equipment and supplies. Multiple companies including The Bendix Corporation, Chrysler, General Motors, Grumman, Northrop, and Honeywell developed plans for Lunar Logistical Systems (LLS) utilizing MOLABS (mobile laboratories) that could carry up to four astronauts for two-week expeditions. These concepts weighed anywhere from three to ten tons and were thirty to forty feet long.

Such vehicles revealed the outsized ambitions of those who saw the space race as the beginning of permanently crewed lunar colonies. As long as these plans remained confined to feasibility studies, they didn't have to be tethered to budgetary constraints, or even practical engineering. But when Robert Seamans slashed NASA's budget in 1967, he nearly put an end to all LLS projects. He'd later joke that,

"Kennedy had tasked us with putting a man on the Moon. Not providing him with a car."

Each moonshot was downsized to a single Saturn rocket. However, if a vehicle could be devised to fit within the existing mission profile, then the project might survive. The only available space was a single triangular quadrant of the lunar module's descent stage. Five feet wide, five feet tall, and three feet deep. Frank Pavlics figured if he could construct a lightweight, foldable rover, then NASA might approve. So he designed and built a one-sixth scale model. His wife sewed the seats, and his seven-year-old son volunteered his G.I. Joe as the vehicle's first test pilot. Pavlics put motors in each wheel, incorporated a remote control, and took the model to Marshall.

Von Braun was on the phone when the vehicle drove through his door. He quickly ended his call. "What have we here!" he exclaimed. After about a half hour discussion, he slammed his fists on the table. "We must do this!"

Chrysler, Grumman, and Bendix all entered the fray with GM—who had partnered with Boeing to take advantage of their aerospace experience—to secure a contract for a seemingly impossible task: construct a five-hundred-pound car that could be stowed in a small triangular wedge and be reliably operated on the surface of the Moon. Everything down to the rover's paint had to be engineered for the most hostile environment. When NASA awarded the contract to Boeing and GM, they entered a seventeen-month sprint to deliver three fully functional, flight-ready space cars.

The final rover looks deceptively simple: a bare-bones jeep-sized dune-buggy equipped with a bulky communications relay, weighing 462 pounds and capable of carrying a payload of roughly 1,000 pounds—twice its weight. Every detail was meticulously scrutinized. Because rubber tires would quickly deteriorate in space and blow up in the vacuum, the vehicle utilized metal wheels consisting of a spun aluminum hub and tires made of zinc-coated steel wires—each of which was hand-woven on a special loom. Loops of titanium bump stops protected the hub and prevented the tire from deforming. Each wheel had a quarter-horsepower electric motor that powered the wheel via a harmonic drive. In case a motor failed, astronauts could remove pins to disengage any wheel, allowing it to spin freely.

Two 36-volt electric batteries powered the vehicle and its communication relay, and yielded a range of 57 miles. The batteries and electronics were passively cooled using a change-of-phase wax and reflective, upward-facing radiating surfaces. The wax would gradually melt like a lit candle as the vehicle operated, and when stopped, astronauts would lift their protective mylar blanket covers and the wax would resolidify, radiating the stored heat out into space.

Originally, a fighter pilot-style joystick controlled the rover, but in test vehicles astronauts complained of wrist fatigue, so a T-shaped handle placed in the center console allowed astronauts easier control.

On Earth, a compass can reliably indicate the cardinal directions for navigation, but this isn't possible on a planet without a consistent magnetic field. Astronauts instead had to rely on an electronic system that continuously recorded direction and distance using a directional gyro and an odometer. The instruments' data tracked the overall direction and distance back to the lunar module (referred to as the LM, pronounced "Lem"). A Sun-shadow device could also give a manual heading based on the direction of the slow-moving Sun.

These are just some of the specifications of an object whose complexity only increased the closer you looked at it. As Earl Swift writes in *Across the Airless Wilds*:

> Basic layout aside, the rover had little in common with any vehicle built in the nearly eighty years of the horseless carriage that preceded, and bore no resemblance to any other 1969 General Motors product, which is essentially what it was. It was called on to cross country that no Earth car would encounter, in conditions that would cripple any terrestrial vehicle instantly: temperatures of minus 250 degrees Fahrenheit in the shade and plus 250 in the sunshine; a surface of clingy, abrasive dust that could foul any moving part; fierce solar radiation; and a constant shower of micrometeoroids smaller than grains of sand but moving faster than bullets. All while wrapped in an airless vacuum.

Such descriptions expose the point where reality and fantasy meet in the lunar rover, a truly alien creation that looked more human than any other spacecraft. The Saturn V was a cathedral, beyond the grasp of the individual. The lunar module floated through the void like an insect or a virus. (Alan Bean called it "the first true spacecraft ever built.") Even the astronauts, fortified in their space suits, looked like extraterrestrials. But the rover is the attainable fantasy, connecting our terrestrial existence to the lunar frontier.

For Bekker, the LRV really was fantasy made real. Not long after the first rover was deployed on the Moon, he received a letter from Jerzy Żuławski's nephew, Juliusz Żuławski, remarking on how the Moon rover reminded him of his uncle's story. Bekker wrote back to Juliusz, and excerpts from the letter were published in 2022 by Adam Żuławski (the author's grandson) in the Polish magazine *Culture.PL.* Bekker relays how delighted he was to receive the letter. "Our paths have crossed in a rather round-about way." He noted that he'd had the illustrated map of the Moon from the 1956 edition of *On the Silver Globe* blown up and displayed, but that the illustrator's drawings of the rover hadn't done Żuławski's vision justice:

> This is why, after translating the description into English, we recreated together with NASA's illustrator a drawing of the vehicle just how it had been described by the author. I have included a photograph of this drawing in this envelope. If you have any corrections or comments, then I am ready to accommodate them. [. . .] I am attaching two photographs of our Lunar Roving Vehicle taken on

> the Moon, along with many heartfelt regards on this very pleasant occasion in which I have unexpectedly made contact with the son of the author of the novel that was always in my parents' library as far back as I can remember.

Between the dreams and fantasies of von Braun, Disney and Bekker, and the realpolitik of the space race and the cold war, it's possible to look at the rover and simply be awed by the audacity of its existence; but it is also a real object, cultivated under real geopolitical pressures—as ephemeral as it is material.

7 DRIVE

The Great Salt Lake is a shallow remnant of the 20,000 square mile, thousand-foot-deep Pleistocene-era Lake Bonneville that filled the basin 18,000 years ago. As its shoreline receded, the paleolake laid down a 90,000-acre salt bed with a five-foot thick crust at its center. When Bill Rishell drove a Pierce Arrow across the flats in 1907, he christened a hallowed ground for racing. Here in 1914, Teddy Tetzlaff set the first Bonneville speed record of 142.8 miles per hour, and Sir Malcolm Campbell breached 300 miles per hour in 1935. To this day, mechanics, engineers, and daredevils from around the world challenge records on the salt, culminating in Bonneville Speed Week each September.

This was my uncle Alan's world. His life's passion involved careening down that salt track at speeds that will make you contemplate the implications of special relativity.

Alan holds a venerated place in the 300-mile per hour club. In 2018, he returned to the Flats for Speed Week as the driver of DRM Racing's long-bodied Blown Fuel Lakester, hoping to break the 306 mile per hour class record. A couple of miles into his run, Alan heard a pop and immediately killed the fuel to the engine. They towed the car back into the pit and discovered that the outside edge of one of its cylinders melted away from the pressure of the high-octane fuel and an aggressive tune-up. But when the team returned in 2020, Alan maxed out the Blown Fuel Lakester at 348.121 miles per hour. The salty dust plume ejected from behind the car made it look like a rocket blasting across the flatland.

8 BARRIERS

In a 1971 issue of *Life* magazine, the Apollo 15 crew appears in a photo standing behind their red, white, and blue Corvette Stingrays. In front of their cars, proudly displayed, is the moon rover.

Corvettes had been associated with astronauts since the early 60s. “Stories abound of these guys hopping into their Corvettes after a night of drinking and setting their 427s and 454s loose along the sandy highways of the Cape, chasing each other and the ever-elusive thrill of speed and danger,” wrote Clayton Seams in *Driving* magazine.

GM sought to make the Corvette the unofficial sportscar of the American astronaut after discovering that they’d been frequenting Jim Rathmann’s Chevrolet dealership for leases and tune-ups. Rathmann, who’d won the Indianapolis 500 in 1960, devised a deal to lease astronauts any new Chevrolet for one dollar. Alan Shepard, Gordon Cooper, Buzz Aldrin, and Gus Grissom all immediately took advantage of the deal. The whole crew of Apollo 12 paid homage to their LM by leasing black and gold Stingrays.

As admittedly alluring as these images are—handsome pilots in sexy cars with sleek space-age lines—undercurrents (and overtones) of nationalism, patriarchy and white supremacy shine through, intentionally and unintentionally. These images, in their unblemished perfection, are reminders of a version of America that has never existed but remains canonized as the American Dream.

Even if you're only looking directly into that yawning nexus where astronauts, rovers, and sportscars converge, you can still find glimmers of a more nuanced story of what could have been. Not far down the road from the Cape, a young woman had set the 1956 Feminine World Land Speed record at Daytona Speed Week. That same year, she broke Cannonball Baker's 40-year record for the transcontinental auto race from New York to Los Angeles. Betty Skelton, known as "The First Lady of Firsts," flew her first plane when she was twelve, and was a champion aerobatic pilot by her early twenties, famed for her "inverted ribbon cut," where she flew her open-cockpit biplane upside down at 150 miles per hour ten feet above ground, cutting a ribbon with her propeller. Skelton set 17 aviation and automobile records and was the first woman to drive a jet car over 300 miles per hour at Bonneville.

In 1959, she became the first woman to undergo the same training as the Mercury Seven astronauts, as part of an article for *Look* magazine. Although she received honorary wings from the Navy upon completion, her entreaties for NASA to consider female astronauts went unheeded.

"I complained that NASA wasn't giving more thought to women pilots [. . .] I wanted very much to fly in the Navy [. . .] but all they would do is laugh when I asked."

Skelton appeared on the 1960 cover of *Look* in a silver space suit, helmet on her knee, in front of a Mercury space capsule, beside the headline, "Should a Girl be First in Space?"

Around that time, Flight Surgeon William Randolph Lovelace II, who helped develop the tests for NASA's male astronauts, began working with Air Force Brigadier General Don Flickinger to conduct an unofficial study with a group of women to see how they would fair undergoing the same tests. More than seven hundred female pilots were screened, and thirteen were ultimately selected as a cohort of First Lady Astronaut Trainees. The trainees were preparing to travel to the Cape when they received word that NASA had cancelled the program when they learned of its existence. The candidates would ultimately take their case to congress in a House Hearing on Gender Discrimination in 1962, during which pilot Jerri Cobb would testify:

> We women pilots who want to be part of the research and participation in space exploration are not trying to join a battle of the sexes. As pilots, we fly and share mutual respect with male pilots in the primarily man's world of aviation. We very well know how to live together in our profession. We seek, only, a place in our nation's space future without discrimination.

In his rebutting testimony, John Glenn would state in no unclear terms: "It's just a fact. The men go off and fight the wars and fly the airplanes. That women are not in this field is just a fact of our social order."

One year later, in 1963 Valentina Teshkova became the Soviet Union's first female cosmonaut—a barrier that wouldn't be broken in the US until Sally Ride's 1983 shuttle flight.

It's ironic that the space program, ostensibly created to advance humankind toward new frontiers and technologies, remained defined by glass ceilings and color barriers. The Kennedy administration, for its part, at the very least understood the symbolic potential of recruiting a Black astronaut. Edward R. Murrow, the journalist and director of Kennedy's Information Agency, wrote to NASA administrator James Webb a few months after Yuri Gagarin's 1961 flight to ask, "Why don't we put the first non-white man in space? If your boys were to enroll and train a qualified Negro and then fly him in whatever vehicle is available, we could retell our whole space effort to the whole non-white world, which is most of it." Soon after, the administration identified US Air Force pilot Ed Dwight as their ideal candidate.

Dwight realized he wanted to fly by the time he was nine, and spent his early childhood cleaning out planes at a local hangar. But as a Black child in Jim Crow Kansas City, Kansas, he doubted it could happen until one day he saw a Black pilot in the newspaper. The man had been shot down in Korea, and Dwight saw him "standing on the wing of a jet, and he was a prisoner of war, and I was like, 'Oh my God, they're letting Black folks fly jets.'"

Dwight passed on a full-ride scholarship to an arts institute to obtain an associate degree in engineering. He then enlisted in the air force in 1953, racking flight hours while developing technical manuals and training pilots on aircraft instrumentation. Despite excelling in his duties, he was told, "country boys wouldn't want to follow me, so I became the number two guy to the squad leader. I wouldn't allow those white guys to outdo me in anything."

When he received his aeronautical degree in 1957, Dwight joined the famed test pilot school headed by Chuck Yeager at Edwards Air Base. Yeager would write in his autobiography that "From the moment we picked our first class, I was caught in a buzz saw of controversy involving a black student. The White House, Congress, and civil rights groups came at me with meat cleavers, and the only way I could save my head was to prove I wasn't a damned bigot."

It turned out he may well have been a damned bigot. Dwight, the only African American of the 26 applicants in a second-phase space training course, was not among the eleven candidates selected to advance. When Air Force Chief of Staff Curtis Lemay told Yeager that the White House wanted an African American participating in astronaut training, he begrudgingly admitted Dwight—and three more white pilots.

Dwight became an overnight icon, appearing in *Ebony*, *Jet*, and other Black publications across the country. He received 1,500 letters a week from fans around the world, mostly just addressed to "Astronaut Dwight, Kansas City, Kansas." But the closest Dwight ever came to space was in a Lockheed F-104 Starfighter, where he flew high enough to observe the curvature of the Earth. "The first time you do this it's like, 'Oh my God, what the hell? Look at this.' You can actually see this beautiful blue layer that the Earth is encased in. It's absolutely stunning."

"To see an Ed Dwight walking across the platform getting into an Apollo capsule would have been mind-boggling in those days," Charles Bolden, the first African American NASA administrator, told the *New York Times* in a profile on Dwight. "It would've had an incredible impact."

Often, when defending the cost of the space program, people have pointed to all of the ways it has advanced technology and our understanding of the universe and Earth's place in it. Alongside this argument, they point to the intangible benefits—the implication of "For all Mankind," and even the words Nixon spoke to Armstrong and Aldrin in his phone call from the oval office to the Moon: "For one priceless moment in the whole history of man all the people on this earth are truly one—one in their pride in what you have done and one in our prayers that you will return safely to earth."

There was, in fact, a spike in public support for Apollo immediately after the first moonwalk, and that narrative is often used as this triumphant capstone to the chaos of the 1960s. But for many, the success only emphasized what the

United States was capable of when it was willing to marshal its resources, which further highlighted national failings on civil rights, women's rights, and economic disparity.

In 2012, on the 50th anniversary of Kennedy's "We choose to go to the Moon" speech, Alexis C. Madrigal wrote a reflection on Columbia sociologist Amitai Etzioni's 1964 book *The Moon-Doggle: Domestic and International Implications of the Space Race*. The book itself is rare and out of print, but Etzioni argues that even the scientific community was divided on the value of the space program—particularly the perceived urgency of manned flight, citing a 1958 report to the President from his Scientific Advisory Committee. "It would not be in the national interest to exploit space science at the cost of weakening our efforts in other scientific endeavors. This need not happen if we plan our national program for space science and technology as part of a balanced effort in all science and technology." Etzioni then shows how, despite the report, the space budget would increase tenfold in the five ensuing years while the total expenditure on research in the US didn't even double. "Of every three dollars spent on research and development in the United States in 1963," Etzioni wrote, "one went for defense, one for space, and the remaining one for all other research purposes, including private industry and medical research."

The criticism of the program, like many criticisms of America, has always been about the disparity between the program's ideals and its practices; about rhetoric versus action. A program conducted "for all mankind" was too myopic to envision a woman or a person of color on the Moon. And a nation that managed to put 12 men on the Moon remained unable to put roofs over the heads of its most vulnerable citizens, or food in the mouths of its most impoverished children. It was these disparities that would compel Gil Scott-Heron to write a poem in response to the Apollo 11 landing called "Whitey on the Moon." "It was inspired," he said, "by some whiteys on the Moon, so I wanna give credit where credit is due":

> A rat done bit my sister Nell
> With whitey on the moon
> Her face and arms began to swell
> And whitey's on the moon

9 ALIENATION

"This is such a big thing, I frankly don't see how you even do it," Alan Bean said of the Saturn V rocket. "Even when you're participating in it, I think it's audacious that you would try. I clearly could never understand it as a crewman, how to make it work. I could only learn how to operate my share of it."

Maybe that's what Timothy Morton was talking about when definining the concept of Hyperobjects. There is no vantage point from which it is possible to view the entirety, whether it's the Saturn V, a rover, the Earth, the Moon, a forest fire, or climate change. My Grandpa Sandy (my father's father) was sent off to Missouri to get "straightened out" after my grandma became pregnant out of wedlock. Sandy took a job pressing holes into sheet metal for Gemini spacecrafts. He would never see that spacecraft in person, save the part he punched the holes into. But the scale of the endeavor mushrooms out to all those people smelting metal, punching holes, riveting, welding, transporting, catering, calculating.

I haven't found Stewart Leroy Sinclair (Sandy's birth name) or Alden Fogliadini's name in any document about Gemini or the LRV. I'm sure there's some yellowed log of

Fogi's work, but more important than that, the rover is a product of collective labor. Not just the hands, industries and institutions that were directly involved, but the entire nation.

The Moon rover is a hyperobject pregnant with infinite meaning. Among that infinity, there's a metaphor for alienation: that capitalist ailment diagnosed by Marx as the process whereby workers are made to feel foreign to the products of their labor. What could be more alienating than seeing that product blasted into space?

One could counter that the collective nature of the endeavor, the idea that reaching the Moon would become part of humanity's legacy, rejects that alienation, but the truth probably lies somewhere between capitalism and Marxism—or perhaps contains both. If we chose to go to the Moon because it was a refutation of communism, then the Apollo landing sites are in some ways monuments to capitalism. But in the mind of a little boy in Ventura, California, they were monuments to the hands that crafted those vessels.

The Moon's prominence makes it impossible for me to forget that legacy. That's part of what distinguishes Apollo. It's still the farthest frontier upon which we've set foot but remains the most immediate frontier in history. We might've looked up to the Moon, but to the Apollo astronauts, anyone who stepped outside on Earth was looking down on them from a blue marble rising in the sky.

Still, we have to contend with the implications of collective ownership—inherited legacies. The right to claim collective credit for Apollo demands a collective culpability for

Hiroshima, Nagasaki, or climate change. The atomic bomb and the internal combustion engine are also products of our collective labor and are as much a part of our national legacy as the Moon rover.

There may come a day, perhaps even in my lifetime, when I will still look up at the Moon and wax poetic about the fact that someone from my family helped craft a car that's resting on its surface. But instead of an optimistic memory, it may become one of a civilization that failed to prevent its own decline, on a planet whose dominant species couldn't marshal that same collective spirit to save themselves. That will be an entirely new form of alienation.

10 MUST MAN EXPLORE

Apollo 15 Commander David Scott and Lunar Module pilot James Irwin descended near Hadley Rille, a deep channel on the edge of Mare Imbrium close to the Apennine Mountains.

"Okay, Houston, the Falcon is on the plain at Hadley. [. . .] Tell those geologists in the back room to get ready because we've really got something for 'em."

Seven hours later, Scott descended the steps of the LM. The sunlight reflected off the astronaut's gleaming gold visor was a stunning contrast to the ghostly monochromatic images of Neil Armstrong.

"As I stand out here in the wonders of the unknown at Hadley," Scott said as he stepped down, "I sort of realize there's a fundamental truth to our nature: Man *must* explore [. . .] and this is exploration at its greatest."

Once Irwin emerged from the module, the two deployed the rover, the corrugated base of its chassis slowly angling out from the descent stage, unfurling like an origami flower, wheels locking into place. Scott and Irwin raised the seats and instrument panel, and then hopped backwards into their seats. "Okay. Out of detent; we're moving," Scott said, uttering the first words spoken while driving a vehicle on the Moon.

The camera mounted on the rover's console is a lower resolution than other cameras used on the mission, giving the footage an old home movie feel. From that perspective, there is no astronaut—it's just the Moon coming toward you and the front fenders bouncing hypnotically in the low gravity. There's something boyish in the astronaut's driving.

"Man, I tell ya, Indi's never seen a driver like this."
"Boy this is so neat."
"Yeow! Whoa!"
"Ol' Barney's really drivin' this beauty."
"Only way to fly, Tony."

There are only two or three minutes of the rover seen from a third-person view, mostly consisting of a test nicknamed the Lunar Grand Prix during Apollo 16. Mission Control wanted footage of the vehicle in motion to assess its performance and functionality. Charlie Duke films as John Young cuts a figure eight across the mare, sometimes sending all four tires airborne.

"That thing is *really* bouncin'!"

The drive is unearthly: a pitch-black sky hangs over a sun-scorched landscape, the ghostly lift and soft landing of the rover, the surreal arc of regolith rooster tails radiating out like the concentric arcs of a wi-fi signal. The dust doesn't linger in the air, because there is no air. It just falls down in accordance with its parabolic trajectory.

Affixed to the Apollo 15 rover was a plaque: "Man's first wheels on the Moon, delivered by Falcon, July 30, 1971." It's a humbler declaration than Apollo 11's, "Here men from the planet earth first set foot on the moon [. . .] we came in peace for all mankind." But the rover's modest statement memorializes both the end of a beginning, and the beginning of an end.

Every mission preceding Apollo 15 had essentially been a dry run, testing concepts and equipment. Armstrong and Aldrin spent just two and a half hours on the lunar surface, venturing no more than 300 feet from the Eagle. Even during Apollo 14, the radius of their excursions didn't exceed half a mile. By contrast, the rover on Apollo 15 would cover more ground in its first thirteen minutes than all previous missions combined. The rover functioned as a storage device, a means of transportation, and a television station outfitted with short- and long-wave antennas and satellite dishes. The whole communications unit could be removed from the LRV and carried with the astronauts when they stepped off to collect a sample or set up an experiment, allowing for more dynamic lunar exploration.

In her book *Timefulness: How Thinking Like a Geologist Can Help Save the World*, Marcia Bjornerud writes that "rocks are not nouns but verbs—visible evidence of processes: a volcanic eruption, the accretion of a coral reef, the growth of a mountain belt. Everywhere one looks, rocks bear witness to events that unfolded over long stretches of time—and are part of those unfolding events themselves." On Earth, the landscape is constantly made new by plate tectonics, weather and biochemical processes. The Moon, by contrast, is a fossilized world whose rocks and features can tell a primeval story.

Harrison "Jack" Schmitt was among the few astronaut-scientists in the rotation for future Moon landings. Since the early Apollo missions, he'd lobbied for a heavier emphasis on science, with little success. To ignite his fellow astronauts' interest in scientific exploration, Schmitt turned to a Caltech geologist to provide astronauts with geological training. Prior to Apollo 15, astronauts visited Arizona and New Mexico to learn how to read the landscape. Simulating missions with space suits and a modified "1-G" rover, they would describe the terrain to CAPCOM, and a geologist unfamiliar with the terrain would rely on the astronauts' descriptions to interpret the findings.

This training would become increasingly critical as NASA began sunsetting Apollo. The near disaster of Apollo 13 and waning public support for the program contributed to the decision to scale back the remaining launches from a planned twenty to seventeen. This left many urging NASA to prioritize scientific objectives on remaining Apollo flights. Questions still remained on how the Moon formed, how old it was, and whether its craters were the result of volcanic forces or celestial bombardments.

The answer to one of these questions would be discovered on Apollo 15's second EVA. Near the edge of Spur Crater,

Scott and Irwin noticed a bright white rock that stood out on the surface.

"Look at that."
"Wow!"
"Almost see twinning in there."
"Guess what we just found. . ."
"I think we found what we came for."
"Crystalline rock, huh?"
"Yes, sir."

The rock contained a large chunk of anorthosite, a piece of the Moon's primordial crust. Dubbed the Genesis rock, the 4.4-billion-year-old sample helped narrow the theoretical age of the Moon, which had ranged among experts from hundreds of millions of years up into the billions.

On Earth, aside from a small finding known as the Jack Hills zircons, there are no other known records of our planet's first 500 million years. "The only extensive source of information for this material—known as the *Hadean* Eon, the 'hidden' or 'hellish time,'" Bjornerud writes, "are samples gathered by astronauts from the Moon. Its familiar, scarred face, with rocks as old as 4.45 billion years blanketed by shattered rock fragments (the lunar *regolith*), attests to a violent regime of relentless impacts as debris left over from the formation of the solar system pummeled the young inner planets."

The anorthosite discovery was an affirmation of the astronaut's geological training, and ultimately contributed to the "giant impact theory" of the Moon, which proposes that the Moon was formed by a collision between Earth and another small planet about the size of Mars. But it also affirmed the utility of the rover, which not only increased each missions' exploratory range, but allowed astronauts to carry complex suites of scientific instruments and facilitated multifaceted and real-time coordination between scientists on Earth and the astronauts on the Moon. To hear the astronauts' excitement at their discoveries is noteworthy in itself—an auditory bridge between the elation of the exploratory phase of early Apollo landings, and the thrill of pure scientific discovery that defined the final missions.

When Apollo 18 was cancelled, NASA came under intense pressure to send a scientist to the Moon. As a result, Schmitt was selected to replace Joe Engle as the LM pilot on Apollo 17, making him the only scientist to fly during Apollo, and one of the last two men on the Moon. Commanding the flight was Gene Cernan, who'd developed a deep passion for geology during training. The two landed in the Taurus-Littrow valley, selected for its access to old highlands material far from Mare Imbrium, and "young" (less than three billion years old) material from volcanic activity. Cernan was responsible for adjusting the antenna on the LRV to maintain transmission to Earth while Schmitt reported on each site's geological features. Meanwhile, scientists in the geology "backroom" on Earth relied on Schmitt's reports to adjust the tasks planned for each site.

During their second EVA, Cernan and Schmitt were surprised by the sight of orange soil near the rim of Shorty Crater. Studies upon return showed that the soil was actually small beads of volcanic glass that had formed more than 3.5 billion years ago—the remnant of molten lava sprayed from a fire fountain high into the Moon's sky.

On Apollo 15 alone, astronauts would cover 17.25 miles in the lunar rover, climbing a mountain as high as Kilimanjaro and arriving at the edge of a canyon a mile wide and a thousand feet deep. The three rovers would travel 56 miles in all, with minimal problems. Two of the three rovers would encounter steering issues, with either the rear-wheel or front-wheel steering being inoperable for a period, but redundant systems made non-issues of these problems. In fact, the most significant malfunction of the missions occurred when a hammer hanging from John Young's shin snagged a rear fender, breaking off its extension during Apollo 17. Without that fender extension, the rover kicked up dust wherever it went, covering the vehicle and the astronauts on their first EVA. The built-up dust, "fine as flower and rough as sandpaper," darkened the surface of the rover, dangerously increasing its temperature. The crew had to constantly stop to brush off the accumulating dust.

Overnight, in a repair reminiscent of the improvised CO_2 scrubber on Apollo 13, flight controllers devised a procedure to tape together four stiff paper maps to form a "replacement fender extension" which they clamped onto the fender. The repair did its job without failing until near the end of the third excursion. It was so successful that the president of the Auto Body Association of America awarded the crew an honorary lifetime membership.

One lead NASA official would remark that the rovers "made the Apollo missions," and one of the astronauts would call the rover "the finest vehicle I've ever driven." Although these machines have been committed firmly to the annals of history, they still occupy a strange place as the only manned vehicular expeditions of another world.

As humans venture further into space on increasingly costly expeditions, it's fair to ask whether or not they are worth the cost, and the risk. During Apollo, the rover functioned as a means of transportation for a geologically trained human. On Mars, rovers have come to function as semi-autonomous robotic geologists, capable of performing complex experiments, albeit less efficiently than a human could. But if efficiency is the primary determinant, might we assume that rovers will eventually catch up?

In his introduction to Anthony Young's *Lunar and Planetary Rovers: The Wheels of Apollo and the Quest for Mars*, David R. Scott affirms his faith in the continued role of astronauts:

> The remarkable advances in computer science and robotics, including software that produces human-like capabilities, seem to indicate that it will not be long before many will say that artificially-intelligent robots of the future should replace the artificially robotic humans

> of the past (picture those somewhat-intelligent Apollo beings of the mid-twentieth century in those old stiff, bulky, heavy pressure suits!). But will robots ever be able to experience the high adventure of exploring the heights of a new frontier? Not really.

So much of what captured the public about Apollo wasn't the science, but the allure of human achievement—the celestial high-wire act. There is a poetry to the humanity of Apollo. James Scott didn't need to test Galileo's theory by dropping a hammer and a feather in a vacuum, but the image endures, revisited in classrooms to teach children about the laws of physics. There is also something charming and haunting about Cernan and Schmitt bouncing around and singing, "I was strolling through the Moon one day. . ."

Then there's all the objects astronauts left on the Moon: the family portrait Charlie Duke laid in the regolith; the Bible Jim Scott left on the rover's control panel. These objects don't prove that humanity *must* explore; but they demonstrate how, when humanity *does* explore, we impart something of ourselves.

In the documentary *For All Mankind*, Jim Schweickart talks about the beauty of spacewalking over the Earth:

> There's a total, complete silence and that beautiful view, and the realization of course that you're going 25,000 miles an hour. You recognize that you're not there because you deserve to be there. You were just lucky. You're the

> representative of humanity at that point in history, having that experience in a sense, for the rest of mankind.

A rover can't have that revelation, but it may ultimately be the conduit of such revelations going forward, and there may be good reason to entrust such an important role to a robot. Humans, after all, don't just bring the beautiful parts of themselves wherever they go. They bring their prejudices and personal shortcomings. A rover can't be mired in disgrace or scandal (at least not of its own design). You can avoid a lot of embarrassment and pain if you remove humans from the equation, but you would also lose those more ineffable insights. Mission planners hadn't placed an image of Earth on the shot list for Apollo 8 when it made the first voyage around the Moon, but when the craft rounded the far side and that blue marble appeared in the void, Bill Anders was inspired to grab a roll of color film and take the photo now known as "Earthrise," often credited for sparking the environmental movement.

There are also the paintings created by Alan Bean, who returned to Earth and wanted to capture the Moon and the Apollo missions as he'd perceived them. He named a painting of Jim Irwin leading David Scott around on the Moon "Conquistadors" because of how much the image reminded him of sixteenth-century explorers. "The conquistadors back in the 1500s came to claim lands and gold and precious gems for the king and queen. But all we came for was knowledge and understanding. A few rocks and a little bit of dust was all we wanted. We carried no weapons. Just tools for digging and measuring. We were space-aged conquistadors, and we truly came in peace for all mankind."

In a self-portrait, Bean is staring into the vastness of space, trying to see if he can find that connection to a higher power that he'd never really been able to find on Earth. The Moon didn't bring him any closer, but the portrait captures that moment, hands raised to the heavens, as an astronaut asks the question that would become the title of the painting: "Is Anybody Out There?"

On the other end of the metaphysical spectrum, Bean painted Jim Irwin kneeling in prayer in front of the rover. Irwin had quoted several biblical passages when he was on the Moon, and became an evangelist when he returned to Earth. He'd tell people that, "The most important thing wasn't that a man walked on the Moon, but that Jesus had walked on Earth." Bean's wife suggested naming the painting, "We came in peace for all mankind."

"That's what we did in Apollo," Bean commented, "and what Jim is demonstrating here is a universal position for coming in peace for all mankind."

11 HEAD-ON COLLISIONS

The driveway up to Fogi's ranch was long and steep. On one side, a barranca separated the ranch from the neighboring property. On the other, avocado trees clung to the slope like they were the only things holding the land in place. The asphalt was run roughshod and tattered on the shoulders from years of baking heat punctuated by downpours and flash floods.

On a May afternoon in 2011, Fogi drove up the driveway and parked his GMC pick-up near his toolshed. He got out of the truck, took to whatever business he needed to tend to that afternoon, and then, for reasons that nobody knows, he got into the passenger seat and put on his seatbelt.

Somehow, he put the car into gear. It started rolling downhill.

Before Jim Irwin completed his final moonwalk, he placed a memorial plaque in the regolith bearing the names of the fourteen astronauts and cosmonauts who died pursuing the Moon. In front of the plaque, he left a small, featureless statuette.

Among the honored dead were the three Apollo 1 astronauts—Gus Grissom, Ed White, and Roger Chaffee—who were immolated in a fire on the launch pad for Apollo 1; and Vladimir Komarov, whose parachute tangled around the Soyuz 1 during its re-entry. US intelligence intercepted Komarov's final radio conversations instructing his wife on how to handle her affairs, suggesting what to do with the children. Moments later, according to the interceptor, "there was just a scream as he died."

There were surely others who died unglamorous deaths from strained hearts and high blood pressure in the name of their countries respective space programs. I like to think that the statuette, in its abstraction, represents those lives, too. As a society, it's worth asking whether the benefits of space exploration justify the risks to life and property. (I'd argue that they do.) But on the individual level, for the person who aspired to orbit the Earth or walk on the Moon, the thought of being "grounded" is a sort of death in itself.

Eugene Shoemaker dreamed of being an astronaut ever since he was a young boy, before Sputnik or Mercury entered the public lexicon. As he came of age with the space race, he studied anything that would improve his chances of being selected as a future astronaut, which ultimately led him to pioneer the field of astrogeology.

Shoemaker was ahead of the curve in understanding that the solar system was more chaotic than it seemed. He deduced from comparisons to the atomic bomb craters at the Nevada Test Site that the great Barringer Crater in Arizona was the result of an asteroid impact. He defended the theory that the Moon's craters were formed by cataclysmic impacts, and that such impacts caused large scale and sudden change on Earth. This was before the acceptance of plate tectonics, continental drift, or the discovery of the Chicxulub impact crater, which proved that dinosaurs had been wiped out by an asteroid. These discoveries showed that the evolution of Earth and its lifeforms wasn't just the result of gradual change over time, but of punctuated equilibrium: sudden changes caused by sudden occurrences.

When NASA came into existence, the organization underwrote Shoemaker's Astrogeological Studies Unit at Menlo Park, California. The organization used newly acquired data from lunar orbiters and telescopes to create a new map of the Moon, and in 1963, Shoemaker became the first director of the Center for Astrogeology as he moved the center to Flagstaff, Arizona, as much for the pristine "seeing" that the high mountain desert afforded, as for the geological diversity of the southwest. Meanwhile, throughout these pursuits, he hadn't relinquished his spacefaring dreams.

Unfortunately, those dreams were dashed amid the move to Arizona, when a diagnosis of Addison's Disease disqualified him from ever becoming an astronaut.

Shoemaker had every right to be devastated, but instead he threw his whole weight behind the astronaut training program. His scientists assessed the Surveyor data, and the branch's lunar maps were used to determine locations for future manned missions. Shoemaker recommended the

Nevada Test Sites to NASA for astronaut training, saying that the fresh craters would provide a more realistic debris-field for scientists and astronauts to understand the Moon.

Between 1963 and 1972, the USGS organized more than 200 geologic field trips for astronauts to sites ranging from the Rio Grande Gorge in New Mexico to Hawaii's volcanoes and Iceland's frozen barrens. At Sunset Crater, a flat expanse of dark cinder surrounding an extinct volcanic blowhole outside Flagstaff, a crew dug 47 holes, stuffed them with fertilizer and dynamite, and blew craters exactly matching their size and placement on the Sea of Tranquility for Apollo 11. These trainings would continue in specially developed training rovers for future missions.

Shoemaker built his career studying the forces that shaped the solar system. He searched the skies for the asteroids and comets that sewed constructive chaos, clumping in ever larger groups to form the first proto-planets, delivering water and complex molecules to the primordial Earth, and occasionally altering the course of history through a sudden and devastating impact.

Starting in 1969, Shoemaker began a systematic search for Earth orbit-crossing asteroids, believing fervently that asteroid strikes were actually common over longer geologic time scales. In 1980, after their children had moved out, Gene encouraged his wife Carolyn to join in the hunt after she expressed a desire to pursue a new interest. Carolyn had no scientific background. She'd studied history, political science, and English literature. When her husband suggested she look through a telescope, she thought to herself, "I've never stayed awake a night in my life." But it gradually came to consume her. "I fell into the program, into the work."

The Shoemakers spent their evenings at observatories logging their findings, reviewing photographic plates, searching for those wandering anomalies that signaled the presence of some new celestial object. Over the course of her career as a planetary astronomer, she discovered a record 32 comets and more than 500 asteroids.

On the night of March 24, 1993, the Shoemakers were at the Palomar Observatory in California with astronomer David Levy when they captured a string of objects orbiting Jupiter. They thought it might just be an anomaly—a lensing effect or some other distortion. But they soon realized that they'd discovered the first active comet orbiting a planet, a realization quickly eclipsed by another, that the comet had passed within Jupiter's Roche limit, the orbital distance at which the planet's tidal forces tore the comet apart. The fragments were destined to collide with the gas giant.

Comet Shoemaker-Levy 9 provided the first opportunity in history to directly observe a comet impact a planet. Every lens from Hubble on down to backyard telescopes was trained on Jupiter in July of 1994. Over six days, the fragments crashed into the gas giant in an event that was televised around the world, leaving scars on the planet's surface that would linger for months, more visible than its great red spot. The largest of the fragments released an energy equivalent of six million megatons of TNT—six times the magnitude of the world's entire nuclear stockpile. The impacts offered the strongest possible argument for investing in planetary defense, while also illuminating Jupiter's role in hoovering up asteroids and comets before they wreak havoc on the inner planets.

The Shoemakers continued their research on craters for the rest of their lives, shifting their focus from looking upwards to considering what would happen on Earth. "Gene had a dream of seeing an asteroid hit the Earth," Carolyn remembered.

Three years after Shoemaker-Levy 9, Gene and Carolyn were on an expedition through the remote Tanami Track, a few hundred kilometers north of Alice Springs, Australia, to meet a friend who planned to help them with some crater mapping. Australia was among their favorite sites, with one of the oldest geological surfaces on the planet. The two lived out of their truck, camping under the stars, searching for craters. As they drove through the Track, "We were just looking off in the distance, talking about how much fun we were having, what we were going to do. Then suddenly, there appeared a Land Rover in front of us, and that was it."

The vehicles crashed head on. Carolyn was badly hurt. "I thought to myself, 'Well, Gene will come around like he always does to rescue me.' So I waited, and I called, and nothing happened."

Overhead, the Comet Hale–Bopp passed perihelion and was moving through the southern hemisphere. In the sky above the Australian outback, it would have been visible to the naked eye—the last comet they'd seen together.

Fogi tried to control the vehicle from the passenger seat as it accelerated downhill, but to no avail. The pick-up rolled into the same barranca where Fogi had scattered Gigi's ashes some years before. He crashed into an oak tree and died of his injuries in the hospital.

The day we visited the ranch after the Thomas Fire, I looked down into that barranca. I remembered my dad dropping Poinsettias, Gigi's favorite flower, into the dry bed every year on Christmas. I thought about Fogi's final moments—how that man who helped GM build a Moon car lost his life in a pick-up built by a GM subsidiary. For a long time, I kind of assumed he'd done it on purpose. He was a cantankerous, old-fashioned eighty-nine-year-old man whose wife had already passed away, whose children had grown, and who'd lived a full life with a legacy from the Earth to the Moon.

My father dismissed that idea out of hand. "It was a freak accident," he told me. I believe him. But I still can't help but suspect that Fogi was ready to inter his ashes back into the soil.

We are creatures of the Earth; wanderers of its lands and seas, rovers in search of its hidden meanings. We toil, we yearn, we live, we die. Even the inclination to leave the Earth, I suspect, harkens back to what Tsiolkovsky said—that this is our cradle, and we can't remain in the cradle forever. But it's only upon leaving the cradle, maturing into whatever we become, that we look back and realize how precious that cradle had been to us, and perhaps we can find comfort in knowing that it too, for most of us, will be our grave.

Only one human being's remains are resting on another planet. On July 31, 1999, a small sampling of Eugene Shoemaker's ashes were placed in a capsule and carried to the Moon on the Lunar Prospector space probe. The brass foil wrapping of Shoemaker's memorial capsule is inscribed with images of Comet Hale–Bopp, the Barringer Meteor Crater, and a quotation from *Romeo and Juliet*:

> And, when he shall die
> Take him and cut him out in little stars
> And he will make the face of heaven so fine
> That all the world will be in love with night
> And pay no worship to the garish sun.

After the Prospector's mission was completed, the probe ran out of fuel and crashed near the Moon's south pole. The impact created its own crater, scattering Shoemaker's ashes among the regolith.

I remember, as a child, being deathly afraid of the Earth being stricken by an Asteroid. That probably has to do with being a young child when the movies *Armageddon* and *Deep Impact* came out. I had recurring nightmares about impacts. What scared me the most was that it seemed unlikely that we would know one was coming—at least not soon enough.

Those fears were mostly grounded in the science of asteroid deflection—the logistical challenge of shifting the course of a planet-killing object in space. In both *Armageddon* and *Deep Impact*, there is this theme that when faced with an inevitable and impending crisis, the government will take action, whether or not they succeed. More recently, director Adam McKay turned to that same scenario in *Don't Look Up* to lambast inaction in the face of climate change, showing how the politics of cynicism, ignorance, and greed come at the cost of our very survival. We are rolling downhill, all gas and no breaks, toward societal collapse.

The average mammal succumbs to extinction in two million years. By that measure, at 300,000 years of age, homo sapiens are relative children—short-sighted, militaristic, nuclear-armed children, who have pushed their home planet to the ecological brink. It's not hard to imagine the impending collapse of civilization, a new dark age, the regression at the heart of that expression, apocryphally attributed to Einstein, that we can't predict what weapons will be used to fight the third world war, but the fourth will be fought with stone spears. In which case, those of us who survive would once again find ourselves wanderers, rovers, on an alien planet.

These are the sorts of things you think about while sifting through what remains after a wildfire. You search the detritus for remnants of what once was. You are a rover in a wasteland, trying to understand *what happened*.

"There's no rhyme or reason to what burns, I guess," Alan said as he ambled toward a mini-fridge in the barn.

But rovers are built and deployed on the premise that there *is* rhyme *and* reason. The Moon must be comprehensible. The history of Mars must be inscribed in its geology. The story of the Moon is in its return samples. Earth *must* be legible, and the solutions to our problems inscribed therein.

"No rhyme or reason." Alan opened the fridge and grabbed three cans of Coors Light. "Like these," he said. "They were in there the whole time, too."

We walked outside the garage, cracked them open and drank.

INTERMEZZO

12 MOONWALKERS

The ground shook when the Proton-K rocket ignited, propelling upward into a clear blue sky in February of 1969. Among the ground crew, the excitement, and tension, was palpable. Loaded onto the Luna space capsule atop that rocket was the culmination of years of work: The Lunokhod, a remote-controlled rover that would embark on the first mobile expedition across the Moon.

Americans had pulled ahead in the space race with the first manned orbit of the Moon during Apollo 8. Meanwhile, the Soviet counterpart to the Saturn V, the N1, had yet to successfully launch. With hope fading of a cosmonaut beating an American astronaut to the Moon, Lunokhod might've helped the Soviets save face. But just moments after liftoff, the Proton-K burst into a cloud of smoke, obliterating the Russian rover.

Just months later, in July, the Eagle descended on the Sea of Tranquility, and Neil Armstrong made humanity's first steps on the Moon.

"My father would say that he personally started believing in communism after Gagarin went into space," an interviewee recalls in Svetlana Alexievich's *Secondhand Time*, her chronicle of the fall of the Soviet Union. I can only imagine how disheartening it felt to see the US flag raised on the Moon—a sight rendered only more traumatic by the Soviet program's tendency to enshrine its failures in secrecy while broadcasting its successes. Had Lunokhod landed that February, Apollo 11 might not have stung so severely. The Russian rover was an engineering marvel. Four and a half feet tall and just over five feet around, the 1,850-pound rover was formed out of a tub-like compartment with a large convex lid, all rolling around on eight wheels with independent suspension, motors and brakes. To operate in a vacuum, a special fluoride-based lubricant was used for the rover's mechanical parts. The electric motors, one in each wheel hub, were enclosed in pressurized containers. The rovers were also equipped with a cone-shaped antenna, a highly directional helical antenna, and television cameras whose lenses were fixed to the front of the tub like a pair of eyes, giving the Lunokhod a particularly *Lost in Space* feel.

Primarily designed to support the Soviet human Moon missions and the planned Zvezda lunar base, Lunokhod was repurposed to be remote-controlled after the successes of Apollo made clear how far behind the Soviets really were. Georgy Babakin, the executive director of the Moscow-based Lavochkin Design Bureau, which specialized in space probes, oversaw the entire production of the Lunokhod and supervised the design of the lunar landing module. Alexander Kemurdzhian designed the rover's chassis at the VNIITransmash Institute in Leningrad, which built, among other vehicles, tanks. Kemurdzhian would bring an imaginative reputation to Lunokhod, experimenting with tank treads, tracked skis and even a bi-pedal walking rover.

When Sergei Korolev, the mastermind of the Soviet program and a survivor of the Gulags, died in January of 1966 from complications during a routine surgery, it became evident that the Soviets would not beat the Americans to the Moon. Korolev was as important to ROSCOSMOS as von Braun had been to NASA, though his identity was largely unknown until the day he was given a state funeral and buried at the base of the Kremlin. Without Korolev, the Soviet manned program was almost certain to flounder, placing more pressure on Lunokhod's success. But sending a rover to the Moon wasn't simple. The Soviets encountered the same challenges and uncertainties as the Americans did with the LRV, but they also had to contend with the additional complications of operating a remote-controlled rover from 250,000 miles away.

Program directors turned to the Red Army to select fourteen officers from a class of forty-five candidates to learn to operate the rovers. "Before us, they had tried to use aviators, tractor drivers, car drivers, and even cyclists," recalls Vyacheslav Dovgan, an officer selected to pilot Lunokhod, in a documentary called *Tank on the Moon*. "They selected men who knew how to assess a situation rapidly, and who could memorize and reproduce the situation immediately."

At a secret hangar in Crimea, the Soviets constructed the Lunodrome, a one-hectare cratered soil bed. Operators would drive the rovers by monitor using a joystick, separated from the vehicles by a black curtain, just as they would during a mission.

As February of 1969 approached, Lunokhod's creators were confident that their robotic vehicle would be the first object to truly explore another world. There was a deep sense of pride, hope, and accomplishment. Even the death of Korolev hadn't diminished the team's drive. If anything, it only made their work more important, and more central to a mission whose success or failure felt comparable to the success or failure of the entire space program.

The Soviets have a "special relationship with death," as another of Alexievich's interviewees describes it. They use euphemistic languages to speak of disappearances, "such as 'arrest,' 'ten years without the right of correspondence,' or 'emigration.'" I imagine that that language was expressing itself quietly in the minds of that ground crew as they processed the explosion of their rocket and the destruction of Lunokhod. And I'm sure that fear only further entrenched itself as Apollo 11 touched down. Regardless, tests for Lunokhod continued in Moscow and Leningrad, often twenty-four hours a day. Workers at the Leningrad office brought in mattresses, sheets, and blankets.

On November 10, 1970, the Soviets launched Lunokhod 1 (the first Lunokhod was renamed Lunokhod 0 in a move that feels almost satirically Soviet). Loaded onto the spacecraft Luna 17, the vessel soft-landed on the Sea of Rains on November 17. As its dual ramps unfurled, Vyacheslav Dovgan "experienced this moment as an amazing burst of emotion. When the landing platform touched down on the Moon, it was tremendous."

Prior to the launch, Babakin's group instructed operators to "make sure the Lunohkod goes down the ramp to the lunar surface and drives at least a few meters. Everyone will then be grateful to you." But Dovgan knew that, as always, there was an unspoken subtext. "We also knew, if the Lunokhod drove only a few dozen meters *and stopped*, we wouldn't be spared."

This mix of fear and elation colored the question of who would drive the rover down the ramp off the lander. Whoever it was would inherit the glory of success, but also risk the consequences of failure. For a moment, the room was still. But then, Gabdulkhay Latypov, one of the operators, broke the silence. "I see the Moon's surface; it's flat. The surface is flat. *And it's beautiful.*"

Latypov took control of the vehicle. Dovgan monitored his vitals as he maneuvered the rover down the ramp. "[He] had a pulse of 140," said Dovgan. "Genya [Gadbulkhay] had never reacted like that before."

At 06:28 UT, the rover rolled onto the Moon. The flight engineer shouted, "The Lunokhod has landed on the surface of the Earth," before quickly correcting himself. "The *Moon*." Everyone broke out in applause. "We could make out the marks left on the surface and it was fantastic," said Dovgan. "A fairytale."

Lunokhod was praised at home and by the international press. The rover was a technological marvel and an unqualified success. Aside from an urban legend that spread among the Soviet Union that the rover was driven by a "KGB Dwarf," Lunokhod's implications were immediately recognized. semi-autonomous rovers would open new possibilities for exploration.

Lunokhod 1, and Lunokhod 2 in 1973, would make significant contributions to cosmology, selenology, and our understanding of Earth. Carrying suites of scientific and navigational equipment including an X-Ray telescope, a radiation detector, retroreflectors and magnetometers, their findings would advance our understanding of the mechanical properties of lunar surface material, observe solar X-rays, and measure local magnetic fields. Combined with the findings of Apollo, the mysterious silver globe had become a familiar world. Questions still remain about the Moon's formation, and frozen water has been detected in the depths of permanently shadowed craters, enticing the next generation to seek out the deposits and discover new questions hidden within its molecular structure. That's the nature of science. The more we discover, the more we realize there is to learn.

Lunokhod operated for 322 Earth days on the Moon, well beyond its 90-day mission. In that time, it traveled 6.5 miles and returned more than 20,000 television images and 206 high-resolution panoramas. Lunokhod 2 likewise outlasted its intended operational period, surviving more than four months and traversing 26 miles of hilly uplands and steep rilles, setting the record for the longest distance traveled on *any* extraterrestrial surface until 2014. The rovers survived the cycles of harsh lunar night and day by stopping occasionally to recharge their electric batteries through the solar array on their hinged lids. During lunar night, the lid would close, and a polonium-210 radioisotope heater kept the internal components at operating temperature.

The Lunokhods would be held up as an exemplar of Soviet technology long after their missions, routinely reproduced in propaganda films until the fall of the Soviet Union. After the collapse, the scientists who worked on Lunokhod began communicating with their American counterparts, who were in awe of what the Soviets had achieved and were eager to see how they had created such robust machines. In many ways, these Soviet rovers helped guide our path to Mars.

Just as Armstrong will remain the only human to first set foot on the Moon, so too does Lunokhod 1 bear the distinction of laying down the first lunar tracks. Its operators, and the people who built it, were sworn to secrecy and relegated to obscurity, but that doesn't mean they occupy a smaller place in history. This wasn't just a technological achievement; but one that proved that a robot was capable of inspiring the same human feelings that one experiences when we see a human walk on the Moon. After all, human or machine, the rover is still *us*.

PART II

Gone is the library, which was burned, its shelves emptied. None of it was permanent. It all fell away.

What parts of us—of all we know, of all we do, of all we are—will escape the same fate? Surely not the rovers, or even those Olympic rings we etched on Mars. Not our current understanding of Mars or its possibilities for life. And, on a planetary timescale, nothing on Earth, for someday the sun will die and swallow it entirely.

—Sarah Stewart Johnson, *The Sirens of Mars*

13 CLOSE ENCOUNTERS

August 27, 2003. The evening sky over Ventura was clear. I sat at the Coopers's kitchen table with their children Glenn and Kari. We'd finished dinner, and Kari dimmed the lights so that Mrs. Cooper could bring out Mr. Cooper's sixty-third birthday cake. As we sang, their rottweiler Brutus rumbled through the final note. Mr. Cooper blew out the candles and we sat around talking and eating until we were ready to go into the backyard for the evening's main event.

Astronomy can be a frustrating hobby. In the eighteenth century, Guillaume Le Gentil sailed from France to Pondicherry in the south of India to observe a transit of Venus. He arrived at a colony in disarray under British occupation. Le Gentil was turned away, missing the transit. He'd have to wait eight years, and travel to Madagascar, where he bided his time mapping Madagascar's coast.

One day, he received a message that the occupation of Pondicherry had been lifted, and he was invited to return. Le Gentil made the voyage back to India and built an observatory in Pondicherry specifically for observing the transit, only for the weather to turn cloudy right at the moment of the planet's passage. The clouds would clear an hour later. Had Le Gentil stayed in Madagascar, he would've had a spectacular view.

A thick foggy marine layer tends to envelop Ventura in the mornings and evenings, but the view was clear as we stepped outside, though that didn't mean the fog wouldn't eventually obscure it. If it did, we would miss a sight that wouldn't occur again for more than two-hundred years.

On average, 140 million miles separate Earth from Mars. Once or twice every fifteen to seventeen years, the distance withers by about two thirds, and Mars appears as a blazing red orb in the sky. On rare occasions, the two planets pass each other when Earth is at its farthest distance from the sun (aphelion), and Mars is at its closest (perihelion), which is what happened on the night of August 27th, when Mars and Earth were on course to pass just 34.6 million miles from each other—the closest approach in nearly 60,000 years.

Even to the naked eye, Mars burned otherworldly bright. Mr. Cooper looked through his telescope, adjusted the focus, and invited us to see the planet for ourselves. I looked through the lens and saw Mars floating in the void, its rust-red surface coming blurrily to life, capped by a white icy pole.

Three-hundred-and-thirty-two miles above our heads, the Hubble Space Telescope turned toward Mars and captured a portrait clear enough to see Olympus Mons, the largest volcano in the Solar System, bulging from its surface, and to make out wispy, blue-tinted carbon dioxide clouds drifting in its frail atmosphere. On earth, a jazz guitarist named Bradley Powell looked down from the night sky and composed a gentle song called "In the Moment (for Earth and Mars)." In the Coopers's backyard, Mr. Cooper continued adjusting his telescope.

Somewhere between Earth and Mars, taking advantage of this close passage, two rovers sailed the void.

14 VITAL SIGNS

The Earth is a living, breathing being adrift between two corpses. Nearer to the sun, a rocky planet with nearly the same mass as Earth and covered with a dense cloudy atmosphere has fallen victim to a runaway greenhouse effect. Though it's believed that it once had liquid water and a habitable climate, the current surface temperature on Venus is hot enough to melt lead, affording Venus the distinction of "Earth's evil twin."

Beyond Earth, a desert world's atmosphere has been stripped away by solar winds, cooling the planet to the point where the water that didn't evaporate into space has receded into the ground or become sequestered in Mars's polar ice caps.

I suspect many people think of planetary exploration as synonymous with the search for extraterrestrial life. Still others might think of it as a search for future habitats for Earth life: places where new colonies can be established, and whole worlds terraformed into new Edens. Elon Musk has suggested using nuclear weapons to vaporize the Martian ice caps and warm the planet (this would require thousands of nuclear weapons and thousands of rockets, and even if the missiles freed all of Mars's sequestered carbon dioxide, the atmospheric pressure would only increase from its current 0.06 percent (that of Earth's) to a whopping 7.5 percent—or about one fourth the pressure at Mt. Everest's summit).

Really, we study the planets to understand ourselves. Even when searching for extraterrestrial life, we are unravelling the mysteries of life on Earth. Whenever we study the geology and morphology of other worlds, we deepen our understanding of Earth. Decoding the greenhouse effect on Venus informed our understanding of climate change, and studying how Mars faltered and withered on the vine helps us understand the precarity of life on Earth.

Every planetary expedition has led to surprising discoveries, from the hundreds of active volcanoes on Jupiter's moon Io, where lava fountains erupt dozens of miles high, to the methane rains that form rivers, lakes, and seas beneath Titan's permanent cloud-cover.

These expeditions have challenged cosmologists' conceptions of what comprises a "Goldilocks," or habitable, zone around a star. Observations of Jovian and Saturnian moons revealed numerous worlds that could support life as we know it. Beneath Europa's icy crust, a vast saltwater ocean may be brimming with the same complex hydrocarbons that gave rise to life on Earth.

We have learned all of this without ever setting foot beyond the Moon. The Mars Perseverance rover deployed the first helicopter, Ingenuity, to fly on another world, and engineers are currently proposing an autonomous submersible called Titan, named for the Jovian moon it is designed to explore. If built, Titan will dive into the liquid hydrogen seas of Kraken Mare, investigating oceanographic phenomena from the chemical composition of the liquid to the sea's tides, wind, waves, bathymetry, and more.

Engineers are even devising rover concepts that can withstand Venus's hellish conditions. Jonathan Saunders, a NASA Jet Propulsion Laboratory (JPL) mechatronics engineer, has taken inspiration from Dutch artist Theo Jansen's wind powered *Strandbeests*, kinetic sculptures that can walk, store energy, and interact with their environment. Videos show these *Strandbeests* walking along sandy beaches like insects or sea creatures, harkening back to a technologically analog era while also evoking organic life.

Venus's atmosphere is so dense that its surface pressure is one-hundred times greater than Earth's, and the 465-degree Celsius temperature on the ground fries most electronics. The longest any man-made object has survived on Venus—the Soviet Venera probe—was 127 minutes.

"For each planetary condition, you need to design a unique application for how you explore," Saunders told *Seeker*. "What I want is a *strandbeest* on another planet." Saunders and his team have been working on their Hybrid Automaton Rover-Venus, known more casually as Harvey, which he describes as combining "a little bit of steampunk with spacecraft to create a clockwork rover." Rather than relying on a motor or battery, Harvey is propelled across the surface by the powerful

Venusian wind. Made out of stainless steel and titanium alloys, the rover crawls across the ground without any guidance. Its silicon carbide and gallium nitride electronics can perform operations under sweltering conditions, although they only have as much computing power as a pocket calculator. Rather than relying on commands from Earth, the rover just observes and reports, cycling through its different instruments. But with luck, Harvey could survive up to 120 days on Venus.

It's not just more cost-effective and practical to send robots to other planets; it may also be more ethical. If there *is* microbial life on Mars, then we should perhaps heed Carl Sagan's advice:

> There will be a time when Mars is all explored; a time after robot aircraft have mapped it from aloft, a time after rovers have combed the surface, a time after samples have been returned safely to Earth, a time after human beings have walked the sands of Mars. What then? What shall we do with Mars?
>
> There are so many examples of human misuse of the Earth that even phrasing this question chills me. If there is life on Mars, I believe we should do nothing with Mars. Mars then belongs to the Martians, even if the Martians are only microbes. The existence of an independent biology on a nearby planet is a treasure beyond assessing, and the preservation of that life must, I think, supersede any other possible use of Mars.

This is why NASA's stringent Planetary Protection categorization system, with strict rules governing missions based on a planet's potential habitability, is so important. The biology of Martian life would likely be so different from Earth life that fears of alien infections isn't very likely, but what is likely, and in fact inevitable in the designs of Elon Musk, would be the destruction of whatever fragile ecosystem sustains that Martian life. If the search for life in the universe is not predicated from the outset with a value for whatever life we find, then the search is destined to be as cataclysmic as any other first contact in human history.

15 PATHFINDERS

I don't remember where I was when Pathfinder launched in December of 1996. I was six. But what I do remember about that year is meeting Mr. Cooper, a retired electrical engineer who lived next to my mother when she was a teenager. Mom babysat the Coopers's son Glenn from time to time, and Mr. Cooper would often demonstrate whatever projects he was working on. To my mother, he was the resident Doc Brown.

Mom wanted me to meet Mr. Cooper ever since the day I walked out of the Boys & Girls Club and sat down in the middle of a busy industrial road—an incident that landed me in a child psychologist's office, and that ultimately drove my mother to conclude that I *really* needed a role model to offer support and encouragement to manage my depression.

She knew from my rock collection and my science books that I wanted to be a scientist. So she reached out to her old neighbor, and on a Sunday evening in the fall of 1997, I ended up on the Coopers's doorstep wearing a white button-down shirt and khaki pants, figuring that meeting a scientist was as hallowed an occasion as church. Mom rang the bell and I heard a growl behind the door. A wispy woman answered, a burly rottweiler at her feet.

"Don't mind Brutus," she said. "He's harmless."

"He's just a dumbhead," came a man's voice from inside.

Entering the house, I saw Mr. Cooper: tall, unkempt white-hair, thick glasses, and a pocket-protector in his breast pocket. All he was missing was the lab coat. He showed me a map of the Moon on the kitchen table, with its craters and mare all labelled, its phases around the rim. Next to it was a box of dull rocks that glowed in a vibrant rainbow when Mr. Cooper shined a fluorescent light on them. Mr. Cooper said I could keep the rocks, the light, and the map, and then we all went into the backyard, where he'd set up a telescope that was larger than I was. He'd made it himself.

Mr. Cooper was a grown child. A child who worked on military projects, testing lasers for Ronald Reagan's Star Wars program and designing other military hardware, sure, but a child no less. As we spoke, he took a laser pointer out of his pocket and projected its beam onto the concrete. Brutus pawed, lapped, and snarled at the light.

"You don't understand lasers, do you, Brutus? You're just a dumbhead."

The Moon wasn't quite full that night, which made for better "seeing." Brightness washes out details, while an angled sun throws features into sharp relief. Mr. Cooper invited me to take a look. As I peered into the telescope, I saw the silkiness of the lunar maria and the black depths of its craters, transforming a flattened disc into a dynamic world.

It was a hauntingly beautiful terrain (Aldrin's "magnificent desolation"), constantly shifting phases but ever-present, even in its absence, in the night sky.

That evening, I'd see the fuzzy lines of Jupiter's stormy atmosphere, the bright points of its four Galilean moons, and the ringed crown of Saturn. I would rove, for the first time, through the solar system.

At the end of the evening, Mom and the Coopers drank coffee and reminisced while I flipped through an issue of *National Geographic*, glimpsing a complex world of people and places whose existence had before gone entirely unknown to me. I was too young to grasp the magazine's colonial heritage, its exoticized or primitivized Other, the whiteness of its lens or the privilege of its editorial focus—I was just swept up in this lush and otherworldly Earth. Each issue exposed a tension between our planet's unremitting splendor and the forces that imperiled it. Strip-mining, conflict, famine, deforestation, wildfires. One headline stuck with me all these years: "Unlocking the Climate Puzzle." I've since tracked down the issue, and it's chilling to revisit a moment when climate change seemed more spectral than imminent, like a shift in the breeze before a hurricane makes landfall. "The world is getting warmer," one caption reads, "and humans are partly responsible [. . .] But many of the pieces of information do not yet fit, and others remain elusive."

I begged mom to buy a subscription. I wanted to be a member of The National Geographic Society—an explorer.

She eventually relented, and in August of 1998, my first issue arrived. On the cover was a little rover named Sojourner. The lead article, “Return to Mars,” was a 3-D feature on the Pathfinder mission. When I put on the red/blue-lensed cardboard glasses that came with the issue, Sojourner came to life, rolling off Pathfinder’s ramp and onto the rocky plain.

It’s possible that I was more enamored with the 3-D pictures than the rover itself, but both seemed so futuristic. History would memorialize Sojourner as the brave little rover that first explored Mars, but it didn’t seem so little to me. I didn’t have any perspective on what was to come. It was my generation, after all, that would come of age with the Mars rovers. We would watch them learn to roll, climb, and even fly as our own bodies figured out how to navigate the Earth.

16 SOJOURNERS

Of all the rovers ever built, "none seem as childlike as Sojourner," Sarah Frostensen says on her NASA podcast *On a Mission*. "The size and weight of a toddler, delivered in the arms of its mother ship the Pathfinder lander." Up until Sojourner's deployment, all of our probes, orbiters, and landers were just ultrasounds and blood tests, only blurrier and less reliable, unable to detect faint signs of life or the nuanced geological history that may have shaped Mars.

Nothing man-made had intentionally rolled across another celestial body since Lunokhod 2. The twin Viking probes NASA sent to Mars in 1975 were complex, multifaceted missions requiring the crafts to enter stable orbits before deploying their landers—a difficult and fuel-intensive maneuver, executed without any clear idea of the terrain or the density of the atmosphere. But those probes would not be able to explore Mars beyond the patch of regolith on which they'd landed.

Still, when the Vikings plunged through that slight atmosphere onto Chryse Planitia and Utopia Planitia, they provided a first glimpse of a tantalizing yet disheartening world. On the one hand, the Marsscape looked startlingly familiar—like the Gobi Desert. On the other, atmospheric and soil tests suggested that the planet was utterly uninhabitable. Ultraviolet radiation, extreme aridity, and the oxidizing effects of the soil led some to conclude that the planet may be self-sterilizing. The prospect of finding evidence of life, past or present, seemed dim, and the cost, complexity, and risk involved increasingly prohibitive. But other scientists, Carl Sagan chief among them, still believed in the possibility of finding evidence of past life, or the conditions for it, on Mars. Viking's findings were too limited to diagnose the habitability of an entire planet. It was like trying to study the Atlantic Ocean from the vantage point of a sea anemone.

While the Soviets had already successfully piloted remote rovers on the Moon, the 1.6-second communication delay between earth and the Moon was nothing compared to the twenty-minute average round-trip signal from Mars. A Martian rover would have to operate largely autonomously, which was difficult with the bulky computer hardware at the time, and far less efficient solar cell technology, which was ruled out as a lightweight power source.

Engineering challenges aside, landing on Mars is just *hard*. To this day, about half of all missions to Mars have failed. In 1971 the Soviet Mars 2 and Mars 3 missions each carried a rover weighing about as much as a cat. The vehicles were mounted on skis and connected to their landers by cables, but unfortunately, Mars 2 crashed upon entry, and Mars 3 landed during a powerful dust storm, failing 110 seconds after touchdown.

By the mid-nineties, Satellites and probes had been launched into space and Voyager had completed its "Grand Tour" of the solar system. But the worlds that so many science fiction writers believed would be adorned with colonial domes like imperial pearls, with rover tracks stringing trade routes between them, had failed to materialize. When Pathfinder was conceived in the mid-90s, it wasn't a return to the multifaceted missions of the '60s and '70s. The organization had amassed a far more muddled record since then. The space shuttle failed to deliver on the promise of regular low-cost flights into low Earth orbit and produced one of NASA's most devastating tragedies when Challenger exploded shortly after launch in 1986, killing all seven crewmembers on live television, including a high school teacher named Christa McAuliffe, who would have been the first civilian in space. And when Hubble was deployed in 1990, its first images revealed an aberration in the mirror that, although only one-50th the thickness of a human hair, resulted in blurry images and necessitated the design of a corrective instrument, installed by astronauts during a spacewalk. Multiple new Hubbles could have been built for the cost of the repair.

These high-profile catastrophes and errors contributed to the development of a new paradigm that would become known as FBC: Faster, Better, Cheaper. Pathfinder was the first significant mission under this approach. Compared to Viking, it would be delivered in half the time, using a team half the size, at a fifteenth of the cost. Unlike Viking's multifaceted mission and complex trajectory, Pathfinder would rocket a lander like a cannonball toward Mars without entering into orbit. This faster approach would subject the lander to a forceful entry, but unlike humans, robots don't turn to putty when subjected to 15 Gs.

The Pathfinder lander was intended to be the primary scientific instrument, and it wasn't until the last minute that a small experimental rover was included as an engineering test, but that rover would end up carrying most of the mission's scientific instrumentation. Originally named the Mars Exploratory Roving System (MERS), budget constraints forced engineers to rely more on commercially available off-the-shelf hardware. The eventual design would be the size and shape of a microwave and weigh about as much as a toddler. Six claw-studded wheels were mounted on a rocker-bogie suspension system, with two camera "eyes" mounted on the front and one on the rear. The rover's main scientific payload, its Alpha particle X-Ray spectrometer, would shoot a laser at Martian rocks to "sniff" their composition. The top of the rover was covered with an array of solar panels, and an additional on-board battery could temporarily power the rover.

On past NASA missions, astronauts named their spacecraft. John Glenn named his capsule *Friendship*, and Apollo landers were variously named *Snoopy*, *Falcon*, and of course, *Eagle*. But Pathfinder's rover wouldn't have a crew to christen her. Instead, NASA outsourced the responsibility to children, reaching out to the Planetary Society in 1996 to organize an essay contest seeking names based on pioneering women. The winning entry was written by Valerie Ambroise from Connecticut, the twelve-year-old daughter of Haitian immigrants. Ambroise wrote about a woman born into slavery as Isabella Baumfree, who was sold away from her parents at the age of nine and would later see her own children sold away from her. When she gained her freedom on July 4, 1827, she joined a religious group that betrayed her, stealing her savings and framing her for the murder of another member. She fled the church and dedicated her life to a path she believed had been laid down before her by God: speaking out against slavery and promoting women's rights, assuming a new name, Sojourner Truth.

"It's only logical," Ambroise wrote, "that the Pathfinder rover be named Sojourner Truth, because she is on a journey to find truths about Mars. The Pathfinder should have strong personalities in order to go under harsh conditions like that on Mars. Truth, while on tours, went under many harsh conditions. Even before, she went under harsh conditions as a slave."

In December of 1996, Pathfinder embarked on a seven-month journey before descending through the Martian atmosphere, a harrowing plunge known as "seven minutes of terror," during which no contact is possible. In that time, the probe evolves from one form to another as it enters the atmosphere at 14,000 miles per hour. The atmospheric drag on the heat shield slows the craft until the parachute can be deployed, reducing the speed further to a brisk 150 miles an hour. Then, the heat shield blows off and the craft drifts down to a height where the airbags can inflate. A Kevlar cord lowers the lander, and then it's cut loose; the craft drops to the surface, bouncing over and over before coming to rest in the dead of Martian night. The team back on earth would have to wait until morning to confirm that all had gone according to plan, but when the sun rose on the first Sol, the lander signaled back that it was in good health. Soon, the dodecahedral petals of its solar array unfurled, upon one of which rested the little rover.

Sojourner's 512 kilobytes of RAM were just enough to store a single image at a time. Its on-board short-wave radio was an over-the-counter device whose 1.2-mile range far exceeded the distance it would travel. With no way for Sojourner to communicate directly with earth, the Pathfinder lander functioned as its relay. The first images revealed that a deflated airbag was blocking the rover's descent, but with a little wiggle of the array, the airbag retracted, and with that, Sojourner finally began to slowly descend the ramp.

Ares Valles is a rocky valley near the Martian equator. Within that valley is a giant outflow channel believed to have been formed by an ancient flood. This was where Pathfinder touched down, surrounded by rocks washed onto the plain from far-off places. The divots, scratches, lacerations, and compositions of each rock were entries in travelogues from places with names like Margaritifer Terra,

Iani Chaos, or Xanthe Terra. When Sojourner made the first chemical analysis ever on Mars, of a rock dubbed "Barnacle Bill," it discovered a story, as Sarah Stewart Johnson writes in *The Sirens of Mars: Searching for Life on Another World*, of a rock that "appeared to have been formed in a relentless cycle of melting, solidifying, and remelting. It was the story of a geologically and hydrologically active world." Rounded pebbles and rounded sockets in rocks all suggested that they'd been tumbled in water. The wind imprinted itself on sand piled into fluted patterns or built up into dunes.

Sojourner was originally designed to remain plugged into the lander, like a dog on a leash. But this would have created needless entanglements. So the cord was removed, allowing free roam for the rover. Even though the rover's communication range limited the distances of its wandering, it would nonetheless live, and eventually die, free. However, Sojourner would have to rely on its own power—a battery that would last for roughly seven sols, and a solar array that provided the rover with seven hours of power per day. Its panels would degrade over time as they became covered with dust and worn down with each charging cycle. As the Martian winter approached, the panels would collect less sunlight, and the temperature would fall until the cold threatened the integrity of the rover's systems. Pathfinder, with its larger batteries and solar arrays, was expected to last a month. Sojourner, even less than that.

Each night, when Sojourner's battery died, the rover would go into a state called "cold sleep," reviving again with the warmth of the sun each day. Engineers called this "Lazarus Mode." No one was certain how long Sojourner would remain operative after the death of its battery. Periodically along its drives, the little rover would stop to send a "heartbeat" message back to the station, just to confirm that it was still operational.

On September 27, Pathfinder's downlink didn't arrive as expected. Mission control tried repeatedly to prompt the lander, to no avail. From that point, it became impossible to know how much longer Sojourner remained active. It's possible that the rover endured long after Pathfinder went down. It was programmed to drive to a more advantageous position if it couldn't communicate with the lander, approaching without getting too close, and circling as it tried to reconnect with its mother ship, like a child nudging the body of a parent she doesn't realize has passed away.

In an interview for *On a Mission*, JPL rover engineer Matt Wallace reflected on how children respond when they learn about Sojourner:

> Whenever we would speak to kids about Sojourner, their first questions were always, "So when's the rover coming home? Who's gonna go get it?" When we talk about the lander/rover, it's almost like it's your kids. [. . .] We portrayed some sort of humanity onto the vehicle. We spoke about it as if it was some kind of living entity, and kids picked up on that, and they were like, "We gotta bring it home, you can't leave it there, it's gonna freeze, it's gonna be cold, it's gonna die."

Some of the mission scientists preferred to imagine that Sojourner circled Pathfinder until it dug itself into a trench, like a mote around its mother ship—just rotating in an endless cycle and never dying out. It's a pretty thought; a childish one.

17 MAD SCIENTISTS

In the third grade, Mr. Cooper and I built a hovercraft. We made a square platform out of plywood with a hole in the center and stuck a leaf blower into the hole. We cut a trash bag into a long, thin strip and glued it around the base to form a skirt. Then we put a lawn chair on the platform. I took a seat, Mr. Cooper plugged in the leaf blower, and the platform began to float on the thin air cushion. Mr. Cooper gave the chair a slight push. I drifted across the yard.

I'd been going to Mr. Cooper's every week for a while. It started on Sunday evenings, observing the night sky through his telescope, but eventually shifted to Tuesdays after school. My friends thought of Mr. Cooper as a mad scientist, which suited him—he *did* have a touch of madness. Not long before I'd met him, he'd gone through a bout with cancer, and he'd once brought seeds with him during treatment and asked the doctors if they would irradiate them. His mind always turned—playfully—toward the diabolical. I remember watching the news with him and Mrs. Cooper once when a flood broke out somewhere. Mr. Cooper took out a sheet of paper and showed how you could add a certain chemical to speed up the waterflow in the upland streams and creeks of a water shed, and then you could throw a bunch of Jell-O mix into the point where those tributaries converged to cause a catastrophic flood.

The hovercraft was a favorite of our creations. Years later, we would build another vehicle by chopping up a bunch of bike frames, welding them to a U-shaped muffler, and attaching a wooden airplane propeller to a motor in the back. Mrs. Cooper saw me ride the propeller trike once before refusing to ever let me on it again. "That propeller's gonna chop his head off, Guy," she told Mr. Cooper.

There was no danger in the hovercraft. It was just a platform floating on a puff of air that could carry a child as far as an extension cord could reach.

One day, Mr. Cooper drove his van to Will Rogers Elementary School and I helped him unload our creation. Our class came out to watch as I explained, at Mr. Cooper's suggestion, how the vehicle worked. I described how the skirt distributed the air evenly, allowing for lift and creating a frictionless vehicle. Then we gave every kid a turn.

Unbeknownst to me, our teacher had called up the *Ventura County Star*, and a reporter was standing in the crowd. She took a picture of me pushing a smiling kid as Mr. Cooper stood beaming in the background, and it ran in the paper the next day.

We made a smaller version of the hovercraft out of balsa wood and a computer fan for that year's science fair, adding a sail to propel it with a blow dryer. It was the fair's most popular exhibit. Unfortunately, we neglected to come up with an experiment. Our poster board explained the science of hovercrafts, but we hadn't posed a question, made a hypothesis, collected and analyzed data, or come to any conclusion. We'd just built something cool.

We took third place, and to this day my mother tells me she couldn't tell which of us was more upset.

That, my friends, is the difference between science and engineering.

18 RUINS

One of the panels on the Mars Perseverance rover is engraved with an homage to that familiar depiction of human evolution, "The March of Progress" or "The Road to Homosapiens," from our early ape-like ancestors to upright humans. The Perseverance etching charts the "evolutionary" course of Mars rovers, from Sojourner to Spirit and Opportunity, then the larger Curiosity and Perseverance. Lingering in the air above Perseverance like an asterisk is Ingenuity, the helicopter that made the first flight on another world in 2022. This engraving is a tribute to all those past missions, linking these rovers to our own evolutionary legacy, further emphasized by a double helix etched inside Perseverance's left wheel, running parallel to a rover tread.

More than anything else, rovers are robotic geologists. On Mars, they analyze rocks, soil and landscapes to determine the planet's geologic history and past, present, or future habitability. But they are also, in and of themselves, signs of life. Imagine an interstellar being arriving on Mars in some not-too-distant future. In a stroke of profound coincidence, on an apparently barren planet, the being happens to descend into the place we call Jezero Crater. Upon a dry riverbed within the crater, it discovers a vehicle, immobile, slowly being entombed by the regolith. As the being examines the vehicle, it notes that it must come from some technologically evolved civilization—but *what* civilization?

It might first ask whether the vehicle was indigenous or alien. It seems disconnected from its environment, but this being knows that civilizations have finite lifespans, and are often reclaimed by their environments when societies collapse. Perhaps these rovers were inter-city messengers, roaming the vast wastes between settlements. Whatever its source, the vehicle's creators have left clues. On the rover's arm are two plates with symbolic writing: "PERSEVERANCE" and "MARS 2020." Each of these are as cryptic and indecipherable to the being as the much smaller script running vertically along the edge of the latter plate: "AONREHMELN1730055." There is also a red, white, and blue rectangle that seems to have some significance, and some prominent cryptic markings: JPL and NASA. Other pictographs on board might be more easily decipherable. Perhaps the being would see the evolutionary plate and make the connection to the vehicle's profile, and from there infer that, at the very least, this vehicle was not alone.

Every rover is a time capsule, as are many, if not all, of the objects we've sent into space. The most famous of these, the Voyager 1, is the first man-made object to travel beyond the heliopause, where the sun's influence wanes and interstellar particles becomes more prominent. On board the probe is the infamous "Sounds of Earth" golden record conceived and arranged by Carl Sagan and Anne Druyan. The record includes greetings from Earthlings, music, a heartbeat, children's laughter, and even a woman's brainwaves. Instructions are etched into the record's casing, along with information about who created it, and where it came from. The wisdom of such messages has long been debated. They might never be found, or deciphered, and if they are, the species clever enough to decode them might use the astronomical coordinates to chart our destruction.

Like Voyager, each Mars rover bears its own cryptic markings. Some of them are obvious to us, like the name nameplates on the Perseverance rover. Others would require explanation even for humans. The sequence AONREHMELN1730055 is similar to a VIN number on a car, relaying information within NASA's designated coding. "REH" denotes that the primary vehicle type is a Rover (R), the power source is Electric (E), and the secondary vehicle is a Helicopter (H). The "3" signifies the third planet from the sun and "0055" means Perseverance is the fifth Mars rover of five attempts.

Other markings tend to reflect the cultural Zeitgeist in which the rovers were conceived. Perseverance, which launched during the Covid-19 pandemic, bears a "unity plaque" depicting an adapted caduceus as a tribute to the world's healthcare workers. The Unity caduceus has only one snake, and the head of the staff is not Hermes's wings, but the globe, around which orbits the spacecraft shepherding

Perseverance toward a small dot in the corner: Mars. The Rock Abrasion Tool (RAT) fixed to the robotic arms on Curiosity and Opportunity were built by Honeybee Robotics. Steve Goroven, the company's founder, was walking to their lower Manhattan office on the morning of September 11, 2001, when he saw a Boeing 767 crash into the World Trade Center. Goroven and his employees watched the towers come down from their roof. For weeks after the attack, access to Honeybee's office was tightly restricted, complicating their ability to meet an already tight deadline. But the idea to memorialize 9/11 on the rover percolated around the office until Goroven eventually reached out to the mayor, whose administration furnished aluminum shards that were fashioned into flag-engraved shielding for the wires on each RAT.

The tribute was kept secret until well after both rovers landed on Mars. Had something happened to them during the journey, it surely would've caused a scandal, though they wouldn't have been the first tributes lost on Mars. Amid the wreckage of the 1999 Mars Polar Lander is a CD-ROM storing the names of a million children for NASA's "Send Your Name to Mars" program. But data on a CD-ROM doesn't illicit the same reaction as a piece of the Twin Towers. I can't deny feeling overwhelmed when my father and I saw a contorted I-beam from one of the towers at the Ronald Reagan Presidential Library. Like a toe of the Buddha or the Shroud of Turin, relics from Ground Zero have been displayed around the world for those who feel inclined to make a pilgrimage, and it makes sense that in the time immediately following 9/11, people at Honeybee felt inclined to cast one of those relics into space. But there must be a subject to revere the object. Should our interstellar traveler's journey lead it to Spirit and Opportunity, there would be no way for it to know that those shieldings had been salvaged from the epicenter of a tragedy.

Every rover is a ruin, a monument to humanity, wandering until its batteries expire or its solar arrays became coated with regolith. Their tracks tell stories, sometimes through literal Morse code: Curiosity's track marks stamp the letters J-P-L into the sand over and over. As our interstellar being wanders the site around the Perseverance lander, it might catch its foot on a small piece of Kevlar. Pulling the cloth from the dust, it will discover the giant supersonic parachute that slowed Perseverance's descent. This could likely encourage the being to consider the rover's interplanetary origins. But that may lead to more questions. Were these strange vehicles all that they sent, and if so, why did they stop?

"The vast distances that separate the stars are providential," Carl Sagan once wrote. "Beings and worlds are quarantined from one another. The quarantine is lifted only for those with sufficient self-knowledge and judgment to have safely traveled from star to star."

As our interstellar traveler unfurls the Perseverance parachute, it may note the orange and white pattern in its sections. During Entry, Descent and Landing (EDL), a camera was pointed at the parachute, and engineers used the markings to determine its orientation. The traveler might be able to infer that purpose, being familiar with advanced avionics, but it wouldn't see that, within each circular row, the arrangement of the sections spells out in binary: "Dare mighty things," a line from a Teddy Roosevelt speech that JPL adopted as its motto. Doubly illegible would be how the phrase echoes the words engraved on the pedestal of a fallen conqueror in an old Earth poem:

> My name is Ozymandias, King of Kings;
> Look on my Works, ye Mighty, and despair!

19 DENIERS

We constructed each floor out of three pieces of cardstock. The middle sheet was folded like an accordion between the top and bottom. We stacked each floor on four toilet paper tubes and ran a straw support beam through each corner of the structure until we had a six-story "building." We took the tower into Mr. Cooper's backyard, set it on the concrete, and poured sand onto each floor to represent the weight of desks, chairs, people, office supplies. Then, Mr. Cooper held a lighter underneath the top floor, close enough to scorch the paper, but not for it to catch fire.

I'd been arguing with a friend about whether a 767 could have brought down the World Trade Center. The conspiratorial video "Loose Change" had just gone viral, and my friend insisted that jet fuel couldn't burn hot enough to melt steel. When I saw Mr. Cooper that Tuesday, I was frustrated.

"Technically, he's right," said Mr. Cooper. "But in every other way, he's wrong."

Over the years, Mr. Cooper and I had fallen into a pattern. Something troubling would happen in the world, I'd try to make sense of it, and he'd devise an experiment. When John F. Kennedy Jr. crashed his plane into the Atlantic Ocean in 1999, I asked Mr. Cooper, who had once been a private pilot, what happened. Kennedy, he said, was an inexperienced pilot, flying blind through a fog bank over Martha's Vineyard. He became disoriented in the fog, and before he realized it, his plane had plummeted into the ocean from 1,100 feet in around thirty seconds, killing Kennedy, his wife Carolyn, and his sister-in-law Lauren Besset on impact.

For that year's science fair, Mr. Cooper and I demonstrated how difficult it is for a pilot to manually control the pitch of a plane. We built a small wind tunnel and a model plane that could swivel up and down on an axis. A motor was attached to a rear elevator flap, and there was a switch that you could use to try to control the plane's vertical pitch. Kids would try to keep the plane level in our "wind tunnel."

(We took first place.)

Through each conversation I had with Mr. Cooper, I was internalizing a mindset that insisted answers could be found if you followed a predefined process: ask a question, form a hypothesis, devise an experiment, examine the results, repeat. The scientific method. In the aftermath of 9/11, a society overloaded with grief and seeking answers became primed to accept even the most dubious explanations, especially if they were packaged in ways that exploited fear and insecurity. I was as susceptible to these conspiracies as anybody else, and the sheer volume of them overwhelmed me. I was introduced to a paranoid lexicon of false flags, body doubles, crisis actors, doctored footage and dubious "expertise." How could *I* refute a structural engineer's knowledge on what could and could not melt steel?

This conspiratorial mindset wasn't the result of 9/11, of course. At a speaking engagement in 2002, Buzz Aldrin was accosted by a "reporter" who asked why Aldrin was getting paid to speak about something he'd never done. "If you've really gone to the moon, swear on this Bible." Aldrin tried to ignore the man, but he kept pushing until, finally, Aldrin punched him in the face.

Conspiracy theories are malleable; they adjust to the moment. There is always a fact to refute every falsehood, but it's impossible to have all the facts at your disposal. The truth is slow, meticulous, and methodical. A lie is quicksilver. But among the most important things I'd learned from Mr. Cooper was that just because *you* can't explain something doesn't mean it's inexplicable. There is a method that can refute madness.

As Mr. Cooper applied the fire to the top floor of our model tower, it gradually began to sink until, finally, it ruptured, and the top floor collapsed onto the next and then the next.

"Everything has its failure point," Mr. Cooper said. "Even steel."

20 LIFEBOATS

The vessel was twenty-two feet long and shaped like a septic tank. The inside of the hull was bare-bones—a couple of support rails, a small off-the-shelf monitor connected to the main computer, an LED light bought from Home Depot, a mat on the floor, and a toilet bucket behind a curtain.

It was the day before Father's Day. Shahzada Dawood was practically giddy about the dive. Hailing from one of the richest families in Pakistan, Dawood paid two hundred and fifty thousand dollars per ticket for him and his son Suleman to board Oceangate's Titan submersible. Dawood's wife had given up her ticket so that Suleman could go. Also on board was Paul-Henri Nargeolet, the seventy-seven-year-old scientist and Titanic expert who'd dived the wreck thirty-seven times before; and Hamish Harding, a British airline executive who'd be making his first dive to the Titanic.

Piloting the ship was Oceangate's CEO, Stockton Rush, who'd set out to disrupt the deep-sea diving world with his novel submersible. Using a pill-shaped hull design that incorporated wound carbon fiber on the cylinder, capped with a titanium half-sphere on either end, the Titan could carry a crew of five. The hulls of other deep-sea submersibles were solid titanium spheres, the ideal shape for evenly distributing pressure, crucial when diving to the 12,500-foot depth where the Titanic rests, and where pressure builds to six tons per square inch.

Shahzada Dawood brought his Nikon camera to photograph the wreck. Suleman brought a Rubik's cube. He intended to set a depth record with it.

On the evening of June 17, the five crewmembers crawled into Titan's hull, the ship's crew bolting the only way in and out behind them. At 4:00 am EST, the vessel began its two-hour descent. An hour-and-a-half later, the mother ship lost contact with the submersible. When Titan failed to surface as scheduled that afternoon, the US Coast Guard was notified. By Monday, US and Canadian ships were swarming the search area. If the crew were still alive, they would have between seventy and ninety hours of oxygen remaining. By Tuesday, with no sign of the vessel, France deployed the Atalante, a ship equipped with a deep-sea diving vessel. But it would take time for the vessel to arrive on the scene.

The search expanded in size and scale on Wednesday. A unified command was established with the US Coast Guard, the US Navy, the Canadian Coast Guard, and Oceangate Expeditions. Remotely operated vehicles (ROVs) investigated the source of sounds detected that may have come from a human-made object. When asked about the cost of such an elaborate search, US Coast Guard Rear Admiral John

Mauger, head of the unified command, declined to offer an estimate.

Days before the Titan began its dive, on the morning of June 14, an activist in Italy named Nawal Soufi was alerted through a special app about a fishing trawler in distress, adrift and overloaded with 750 migrants as it crossed the Mediterranean. Soufi sent the message and the ship's GPS coordinates to the Italian, Greek, and Maltese authorities. Early that afternoon, a surveillance aircraft from the European Border and Coast Guard Agency spotted the vessel and alerted the Greek authorities, who confirmed that they'd already been notified of the vessel on Tuesday morning.

A couple of hours later, activists received the first call from the boat's passengers saying they could not survive the night. Greek authorities said they'd established contact with someone from the vessel, but that they didn't request any assistance. By 3:35pm, a Greek Coast Guard helicopter located the trawler and released a photo showing a deck packed with people, some with outstretched hands. At 5:10, at the request of the Greek authorities, the Lucky Sailor tanker brought food and water to the trawler, but authorities claimed that people on the boat were "very hesitant to receive assistance" and were shouting that "they wanted to go to Italy." The trawler was later convinced to receive supplies. At 5:20pm, passengers called Alarm Phone saying that the boat wasn't moving, and that the captain had fled on a smaller boat. Meanwhile, the Greeks claimed that they'd observed the craft "sailing on a steady course." Soufi would report more calls from the vessel that evening saying that six people had died and two were sick.

The ensuing hours are mired in confusion. The Greeks claimed that the passengers were not seeking help and desired to go to Italy, while activists insisted that passengers had been desperate and that the Greeks were attempting to tow the vessel into Italian waters to absolve themselves of responsibility. It's certain that by 11:00 pm the trawler began rocking as its passengers tried to catch water bottles from another vessel. Ropes were tied to the ship, further destabilizing it and sewing panic. The last message Alarm Phone received from the ship was a desperate passenger: "Hello, my friend. [. . .] The ship you send is—" the message ended there.

The vessel began to rock violently until it finally capsized. Within ten to fifteen minutes, the ship had sunk, killing all but 104 of its passengers.

As the no-expense-spared Oceangate search played out with real-time coverage as elaborate as the rescue operation itself, the disparity in the responses became hard to ignore. Not since the sinking of the Titanic itself had a maritime disaster so clearly illuminated the grotesque architecture of class to a global audience.

Greek authorities had already been exposed loading migrants onto a boat in the middle of the ocean and shoving them off to sea to either wash up on some other country's shore or succumb to the waves; and the Pakistani, Egyptian, Syrian, and Palestinian governments, whose citizens boarded the ship, were quick to attack the rescue effort while failing to accept culpability for the conditions that drove out their citizens and allowed the boat to set sail in the first place. The Greeks and other Mediterranean authorities defended their own actions by crafting a favorable narrative. They also stood by a policy of dissuading providing aid to distressed migrant ships to discourage more migrants from attempting the crossing.

That same logic, it seems, did not apply to the passengers on board the Titan. No one suggested that such rescue efforts only encourage billionaires to risk their lives. By the end of the day on Wednesday, an ROV discovered wreckage from the Titan on the sea floor about 1,600 feet from the Titanic's bow. It was then revealed that the US Coast Guard had detected a sonar signal that aligned with the submersible's last known contact, strongly suggesting that the hull had imploded. The wreckage confirmed that belief. The vessel violently collapsed in a fraction of a second. The Titan's crew were dead before they could know anything had gone wrong.

This would not have been the experience of the migrants trapped below-deck on that Mediterranean trawler, or for the people above deck who died of exposure, dehydration, or drowning after the ship capsized. Even in death, there was a marked disparity in suffering.

Before the Titan's implosion, no one had ever been killed in a deep-sea submersible. Earlier ventures like Alvin were expensive endeavors funded by the US Navy and the prestigious Woods Hole Oceanographic Institute, and had been built by General Motors in a process requiring years of development and testing. Among the cardinal engineering principles that the Alvin team followed was not to fix what wasn't broken. A conservative approach to engineering limited variables and risk. Innovations demanded meticulous testing, including rigorous stress tests to see how a vessel would handle enormous pressures. In the end, a vessel would be "classed," rating it for safety, depth, and other factors. Even privately funded vessels such as James Cameron's Deepsea Challenger submersible endured rigorous testing. It's true that Cameron's vessel wasn't classed, but when asked for comment on the Titan, he noted that the only life he risked with his submersible was his own.

The quirks of maritime law leave the regulations surrounding deep-sea submersibles in a murky place. Companies like Space-X, Blue Origin, and Virgin Galactic are subject to far more regulatory scrutiny than Oceangate. Hamish Harding, who lost his life aboard the Titan, had flown on a Blue Origin suborbital flight just the previous year—a trip with a ticket cost between two and three hundred thousand dollars. And for that flight, he signed a similarly catastrophic liability waiver. But those flights are regulated by the Federal Aviation Authority, and are subject to regular scrutiny. Aside from Crista McCauliffe, no private citizen has died during spaceflight. But the risk is very real, and in fact, the death of the Challenger crew forced NASA to come to a reckoning that resonates with Oceangate. The space shuttle was always an experimental vessel. For its operators, safety was paramount, but the Reagan administration and NASA's portrayal of the shuttle as safe enough to transport a school teacher tattered the agency's reputation. Space travel is, and always will be, dangerous. It seems almost inevitable that with an increasing number of private space flights, one of them will eventually fail. If and when that happens, it might be easy enough to say that if the richest people want to spend their money risking their lives, that's their prerogative. Let them freeze to death waiting to summit Mt. Everest, or blow up as they slip the surly bonds of Earth, or become compressed into oblivion miles below the ocean. Surely, they have earned that privilege.

But Titan reveals the fallacy of that reasoning. When the wealthy are imperiled, small battalions are marshalled for the search. Heaven and Earth were reordered to look for five wealthy adventurers, whose odds of survival were virtually zero, while hundreds of migrants drowned in plain sight.

A similar disparity was apparent when I returned to California after the Thomas Fire. Of the thousands of homes that burned, the most susceptible to fire were the hillside citadels of wealth. Those homes, farther from the center of town, required additional water lines and firefighters, thinning vital resources. On this swiftly changing planet, there is no question that the wealthiest among us can basically live a hermetically sealed existence, protected from climate extremes by their freedom of mobility, their access to capital and influence, and even, in the minds of some, the possibility of leaving the planet entirely.

As I followed the search for the Titan, I thought of Jan Korecki, and his regrets and realizations about his fool's errand. I wondered what might have been going through the Titan divers' minds as they awaited rescue. For Nargeolet, this might've been the death he expected, albeit an ignominious, perhaps even humiliating, permutation of such an end. I wondered if Stockton Rush would be cognizant of the place he'd earned among the explorers whose hubris cost his crew their lives. For Harding, I suspect that his age and his experience would have allowed for some resignation. But for the Dawads, I can't help but imagine the fear, the regret, the sense of betrayal they would surely have felt toward the man who had sold them a dangerous illusion. I come back to Korecki, repeating to himself, "I will be dying with these people as a voluntary sacrifice of that all-powerful desire for knowledge."

Of course, the divers actually thought nothing of their fate. They didn't have the chance.

Rush said he believed in the value of democratizing deep-sea exploration, in the same way that Bezos and Musk use the overview effect as a selling point. If more people, especially those of influence and means, can see the Earth's cosmic fragility, then that perspective can influence their sense of purpose after they land. William Shatner invoked this experience when he completed his Blue Origin flight. "When I was there, everything I thought might be clever to say went out the window."

"Everything else just stood still for a moment," he added, on the verge of tears, if not outright weeping. "I was overwhelmed with the experience, with the sensation of looking at death and looking at life. It's become a cliché of how we need to take care of the planet, but it's so fragile."

I believe that Shatner's experience was sincere, but I also believe that he'd been primed for it. For years, Shatner has been a space exploration advocate. In my opinion, he had his overview epiphany long ago.

I can't read the hearts and minds of Bezos or Branson, or whether their flights altered their perception and sense of purpose. All I can say for sure is that none of these billionaires seem to be acting all that different post-flight. Elon Musk has never gone to space, but I don't suspect his takeover of Twitter (now X) would have been executed any more humanely, with any more empathy or recognition of his responsibility toward his fellow humans, were he to have spent a few hours in orbit.

It seems to be a pattern among tycoons to look enviously at the pioneers who first conquered the frontier, and to want to replicate those accomplishments, experiencing first-hand those transcendent moments. If you read enough narratives from the great explorers, you may find that the cliché is true. The thing these tourists are looking for wasn't in the destination, but in the journey. More importantly, true exploration is reaching into the unknown, and yearning to answer a question that, for all time, has remained unanswered.

Robert Ballard wasn't 12,500 feet beneath the Atlantic when he first discovered the wreck of the Titanic. He was aboard the Knorr, a US Army research ship, from which he deployed *Argo*, a 16-foot submersible sled equipped with a remote-controlled camera. This new technology, which Ballard called "telepresence," allowed for dives that could last nearly indefinitely, without endangering a crew. On September 1, 1985, the live feed from *Argo* revealed the Titanic to human beings for the first time since it sank in 1912. In a lecture for the Inner Space Institute at the University of Rhode Island, Ballard talked about how controlling *Argo* through the monitor felt as if he was there at the Titanic. To him, there was little difference in the experience of exploring the wreck in person and exploring it through his ROV. *Argo* was a more efficient explorer, maneuvering through portholes to the ship's interior, which would have endangered a crewed submersible. Ballard would dive down to the Titanic himself a year later in 1986, and on numerous occasions beyond that, but these would be legitimate scientific investigations backed by renowned institutions, and his stance on the ROVs would never change.

The direct descendants of *Argo* would be deployed in the search for Oceangate's Titan, trawling the sea floor and the water column above it. As the wreckage, along with unspecified human remains, surfaced, the relatives of those who were lost on the Mediterranean migrant trawler pleaded for the recovery of *their* loved ones. Adil Hussain, whose brother Matloob went down with the ship, told *Reuters* that he'd left the door of his Athens home open for Matloob. He planned to keep it open until the body was returned. "They must take out the people who are inside. If they are dead, take them out [. . .] We will sell our houses, we will borrow money, if the state can't. Just give me the body." The Greeks made clear that such a recovery was unlikely. The vessel sank in waters more than 16,000 feet deep; deeper, even, than the Titanic.

There are ways to recover the fishing trawler from the bottom of the ocean, but those methods would be insanely expensive. The point here is not the practicality of such an effort; but that cost was not in question during the Oceangate search. There was no pleading, no weeping to coerce a government into action.

When Apollo astronauts talked about the overview effect, they described a planet adrift in a vast emptiness, like a ship alone out on the ocean. In that cosmic ocean, whole planets and whatever civilizations conquered them are often destroyed, like sinking ships. I wonder how many of those civilizations gave rise to tiny, insignificant beings who felt they had more of a right to life and freedom than any other beings on their planet.

I find myself thinking of Elon Musk's Starship as a lifeboat, ferrying privileged humans off of a doomed planet, like the wealthy first-class passengers whose proximity to lifeboats and understanding of English inherently improved their chances of survival on the Titanic. It reminds me of Kim Stanley Robinson's indictment of the current global economic system that allowed a sort of planetary genocidal profiteering, writing in *The Ministry for the Future* of:

> the many ingenious ways that finance had found to short civilization, thus in the process shifting most of the surplus value created in the last four decades to the richest two percent of the population, making those few so rich that they could imagine surviving the crash of civilization, they and their descendants living on into some poorly imagined gated-community post-apocalypse in which servants and food and fuel and games would still be available to them. No way, she said to the bankers; not a chance that would happen. Shorting civilization and imagining living on in some fortress island of the mind was another fantasy of escape, one of many that rich people entertained, as ridiculous as retreating to Mars.

If civilization collapses, I wonder if those humans living on Mars, should that "ridiculous" fantasy become manifest, might return to the ruins of Earth to see what once was. Through generations on Mars, they'd likely need to send rovers to manage among the gravity that might otherwise overwhelm them like a dense ocean. Or perhaps the wealthiest among the Martians will fashion their own MOLAB rovers and exoskeletons to be the first Martians to return to Earth in spite of the danger. Whatever lens they look through, they might marvel at the fallen civilization, its crumbling buildings and the life that still manages to endure. Maybe they'll even argue about the ethics of excavating relics from a grave.

21 MENTORS

Mr. Cooper swam around the Ventura Pier multiple times a week. Once a week, he'd ride his bike up the steep trail to Ojai and stop at Bart's Books to peruse their science section. He walked his dogs along trails in the foothills, always carrying a trash bag and a grabber to pick up any garbage he saw. He was a volunteer swim coach for the Special Olympics. I can still hear the tune he whistled while he worked. He had surplus hardware from guided missiles in his garage. He tried to invent devices that would alert the world when tsunamis occurred, and alert him and his wife when his mother-in-law leaned too far back in her chair. He traveled the world with his wife and sent me postcards from everywhere he went. He flew planes. He imagined deflecting asteroids by painting them white and pushing them with light. His library smelled like old books, machine grease, coffee, and dog. He pronounced robot "robutt." He had twenty-four patents to his name. At seventy-six, the doctor said he had the heart of a twenty-four-year-old. He died of a brain aneurysm a few weeks later. I still picture him in his garage, rummaging through Folgers tins for nuts and bolts, making magic from spare parts.

He wondered about everything. He wandered the world. He was a rover.

22 THE ROVER AT THE END OF THE WORLD

I was a high school senior when *Wall-E* was released, and I can still remember the movie's opening sequence, starting with the universe's spiral galaxies, nebulae and star clusters filling the screen in Disney/Pixar splendor. Meanwhile, the song "Put on Your Sunday Clothes" from *Hello, Dolly* dreamily projects a world "full of shine and full of sparkle."

It was awful. Tommy Tune's voice was so cheerfully nasally, and in Disney's hands all of this sidereal splendor primed me for distrust.

But as the sequence ventured through space and arrived in our celestial neighborhood, I was surprised at the sight of Earth. It wasn't the Earth envisioned by Walt Disney, Tomorrowland, Wernher von Braun, Elon Musk, or Konstantin Tsiolkovsky. This was a dead Earth, its hazy atmosphere a sickly Venusian yellow, obscured by a cloud of inoperable satellites and space junk. Descending through the cloud tops, the world appears as an arid planet whose mountain peaks and rolling hills are all piles of refuse. Wind tunnels and power plants sprout like dandelions and mushrooms from the landscape before a metropolis arises, its ruined skyscrapers interspersed and overshadowed by towers of garbage like stacks of compacted cars in a salvage yard. The question of who is compacting and stacking all those cubes is soon answered when we swoop down a winding path through the societal garbage heap to a lone laborer diligently collecting rubbish, shoveling it into his mechanical carapace, applying pressure, and spitting out another cube to add to his latest tower.

I wouldn't call a garbage truck or a trash compactor a rover. The singularity of their purpose and function seems too limited to categorize them as anything but heavy machinery. And Wall-E—the Waste Allocation Load Lifter-Earth class—definitely qualifies as heavy machinery, designed with the Sisyphean "directive" to manage civilization's cumulative waste. His diminutive, rusted and battered appearance only underscores the outsized scale of the problem he's been manufactured to redress, though not alone. Wall-E rolls by a Soviet-style billboard showing a line of Wall-E's confidently charging into their mission. *Working to Dig You Out*, the sign promises.

But if all that makes Wall-E a machine, then it's what makes Wall-E feel human that defines him, for me, as a rover. It's the discarded Rubik's cube he stores in his lunch pail, the cockroach he befriends, and his Arielle-esque collection of gadgets and gizmos aplenty. Disconnected from the implications of his dystopia, Wall-E finds, and more importantly *seeks*, beautiful remnants. At the same time, as he persists in his futile mission to reshuffle human waste and combat entropy by cannibalizing the rusted remnants of other Wall-Es for his own survival, he seems to inevitably reflect the drudgery of persistence amid ecological catastrophe.

Passing some sort of monolithic transit hub, an automated video advertisement begins to play. *Too much garbage in your face? There's plenty of space out in space. BNL's Starliners leaving each day. We'll clean up the mess while you're away.* The BNL (Buy N Large) CEO steps onscreen to make his final pitch: *At BNL, Space is the final fun-tier!*

This is the uninhabitable Earth. The Earth destroyed not by asteroids or nuclear war, but rampant consumerism: terra-deformation. Rearranging the deck chairs on the Titanic. But in the optimistic spirit of the Buy N Large corporation, technology will dig us out, without the corporation ever having to acknowledge that they were the ones who buried us.

Rovers, as I wrote at the beginning of this book, are machines designed to transport us through inhospitable environments, or to travel there in our place. The Moon, Mars, Venus, the oceans of Titan and Europa. Increasingly, they can reach into the most hostile corners of Earth—the depths of our own oceans, deep within caves. As computers become more powerful and robots more sophisticated, more of the world is within a rover's reach than ever before.

Yet simultaneously, more of the world is receding from habitability. In the summer of 2023, as I began wrapping up this book, Phoenix, Arizona endured a sweltering and record-breaking thirty-one straight days of temperatures above 110-degrees Fahrenheit. The city created the country's first department of heat management, advising families to limit their time outside to early in the morning or late in the evening. Worldwide, June of 2023 appears to have been the hottest month in Earth's history, and July and August aren't looking to get any cooler—nor any summer in the years to come. Those rising temperatures are warming the oceans, increasing their acidity and contributing to dangerous coral bleaching and death—endangering one of Earth's most vital ecosystems. As the polar icecaps melt, island nations and coastal communities are either desperately attempting to bolster their infrastructure or abandoning their homelands entirely. Already, the world is seeing the first waves of climate migration, as millions flee fires, floods, and droughts, while millions more suffer with no respite.

Scientists warn of a 2-degree tipping point that will accelerate the already terrifying effects of climate change to something of a more biblical proportion: a total climate meltdown. Ocean currents and the jet stream may stall or reverse course, the Arctic ice cap could completely melt, gigatons of methane could be released from thawed permafrost, and heatwaves could grow so hot and last for so long that they could cause the death of millions.

Humans are still coming to terms with our ability to render large swaths of the planet uninhabitable. Despite an increased consciousness of the scale of the universe and our insignificance within it, in our daily lives, Earth is still an all-encompassing, impossible to fully perceive hyperobject. We are motes of dust on that behemoth. And ironically, the ability to understand the universe, the Earth, and even the weather, may in some ways undermine the imperative to shift away from fossil fuels. Through solar panels, wind turbines, geothermal plants, electric cars, and exotic ideas like geo-engineering and carbon sequestration, some might feel assured that the world will rally at the last minute to avert catastrophe. That's the scenario envisioned in an early episode of *Futurama:*

> Fortunately, our handsomest politicians came up with a cheap, last-minute way to combat global warming. Ever since 2063 we simply drop a giant ice cube into the ocean every now and then [. . .] Of course, since the greenhouse gases are still building up, it takes more and more ice each time. Thus solving the problem once and for all.

Politically, we certainly exist in an ecosystem underpinned by cheap, last-minute solutions, and I wouldn't refute either the necessity or the potential for technological advances and shifts in energy production to help address the climate crisis. But I also believe that there will come a time when we have to reckon with the aspects of ourselves that lie at the root of the problem.

When I think about this, my mind turns to Chernobyl. The explosion at Chernobyl's Reactor 4 in April of 1986 marked the world's first peace-time nuclear disaster. Initially treated like a normal explosion, fire fighters were the first on the scene, not realizing that the rubble they were walking over included radioactive fragments of the graphite rods inserted into the reactor core to slow the reaction process. Nor did they understand that the water they sprayed onto the open roof of the reactor was just boiling off, contributing to the radioactive cloud drifting across Europe. Those firefighters, along with some of the plant's operators, would be among the small number of deaths the Soviet Union would directly attribute to the disaster.

To quell the initial fire, helicopters flew above the reactor, dropping sand down into the core. Once the fire abated, robotic response vehicles were considered, including an East German-built remote bulldozer, but weighing in at dozens of tons, it would be too heavy to operate on the partially collapsed reactor building roof. "The robotic devices we had, be it our own or acquired from abroad, turned out to be practically useless in those conditions," Valery Legasov, an inorganic chemist who was assigned to respond to the disaster, said in his tape-recorded memoirs shortly before his death. Among those to respond to the explosion, Legasov realized he was dying of acute radiation sickness months after the disaster and chose to hang himself. "Say, even if a robot had sufficiently reliable electronics, it could not overcome obstacles, that were the result of a large amount of wreckage of the fourth block, and stopped. This was the reason they were unusable. If, however, the researchers received robots that had good all-terrain travel capability in the most difficult conditions, then their electronics would fail because of the high gamma radiation and they also stopped."

This was a scenario where robots couldn't be relied on to respond. Instead, human laborers, who would become known as "bio-robots" and "liquidators" were called to action. Working in 90-second shifts, liquidators donned improvised lead-lined materials, and then ran onto the roof, scooped a shovelful of debris, and threw it into the reactor core.

The footage of these liquidators looks like something from another planet. Cloaked from head to toe in alien gear, they run out onto the roof, perform their task and quickly run back. Some can actually be seen picking up chunks of graphite with their hands.

It's estimated that somewhere around 500,000 liquidators were called on to respond to the Chernobyl disaster, with many later succumbing to cancer, or birthing children with defects. Miners dug beneath the reactor to install a protective shielding to prevent the core from melting down, leeching into the water supply, or causing an even larger explosion. Soldiers roamed the surrounding city shooting dogs to prevent them from roaming into uncontaminated areas.

But there were limits to what humans could do, too. Some debris was simply too heavy for people to move. To address the problem, leaders in the Kremlin summoned one of their leading experts in experimental vehicles, Alexander Kemurdzhian, to Moscow. Kemurdzhian would have three months to take the expertise gained from building Lunokhods, Marsokhods and their broad array of planetokhods to construct a remote-controlled vehicle that could clear the deadly debris projected by the explosion onto the roof of reactor 3.

Kemurdzhian returned to his team at Transmash, presented the scenario, and asked them, "How can we help them? What can we do?"

Immediately, the team began working on a miniature version of the Lunokhod called the STR-1. Working around the clock, it took them one week to design the rover and another week to almost completely build it. The result was a fully automated titanium bulldozer, with notched wheels cut from solid blocks of metal alloy to allow it to traverse the power plant's mangled tar roof. The rover was ready to ship on July 15th. The radioactive environment meant that its operators would have to be nearby, and since the engineers were the only people who knew how to operate it, they'd have to do it themselves.

Despite the deadly risk, the engineers traveled to Chernobyl, and two STR-1s were dropped onto the roof by helicopter. Pavel Sologub led the first team, operating the vehicle throughout the entire month of August. Mikhail Malenkov led the second team and spent thirty days on the contaminated roof. "I agreed to go enthusiastically. I understood where I was going. I took risks, but I knew what to do, and where we could or couldn't go and what was feasible or not."

The STR-1s proved useful for clearing debris, but even these radiation-hardened adaptions of space rovers couldn't withstand the environment on the roof of the reactor for long. Both of the STR-1s would eventually fail, and human liquidators would once again cloak themselves in lead to clear the rooftop. Still, it's estimated that the amount of debris removed by the STR-1s was enough to offset the need for thousands of liquidators.

This is part of the reason that some called Chernobyl a crisis that only the Soviet Union could have created, but that only the Soviet Union could have solved. The errors that led to the meltdown were the product of bureaucratic processes and a fear of reporting errors, all of which magnified to the point where the reactor exploded. But it's been suggested that only the Soviet Union would be willing to send 500,000 of its own people into a radioactive hot zone to do whatever it took to solve the problem.

Part of me is inclined to believe this line of reasoning, but then I step back, and again I turn to the climate crisis, and see the same Kafkaesque logic in the words of climate deniers and fossil fuel industry representatives. The same companies lobbying to preserve oil subsidies and curb regulations are running commercials about their investments in clean energy; and politicians whose constituents are suffering in so-called "sacrifice zones" where harmful industries are concentrated, or whose lives are being disrupted by natural disasters are actively denying the obvious causal relationship between increased greenhouse gases and more frequent and powerful natural disasters. We've demonstrated our own willingness to sacrifice lives in the face of climate change—but that sacrifice has had nothing to do with combatting the issue, and everything to do with a high tolerance for collateral damage, so long as the economic system continues to incentivize negligent behavior and enrich a small and hermetically sealed off sector of the population.

Chernobyl remains to this day an exclusion zone, sparsely populated by a few stubborn survivors of the disaster who refused to evacuate, and visited by disaster tourists who marvel at the ruins. But the disaster has also turned the site into an unintentional greenspace, and researchers have been tracking the feral offspring of the dogs that survived the initial slaughter, along with the insects, trees, wolves and other species that have filled the ecological niche, unbothered by humans, but not unscathed by radiation. The site offers a strange sort of hope, and a reminder that nature is more resilient than our ability to strangle it.

It brings me back to that little plant Wall-E discovers amid the rubble—the spark of hope that Eve arrives on the planet to find. In fact, Chernobyl was one of the sites that the creators of Wall-E turned to for inspiration, not just for its history with apocalyptic robots and its ruinous city, but for the resilience of nature.

When Eve's mothership returns her to the human lifeboat/cruise ship Axiom, Wall-E clings to the outside of the rocket for dear life, transcending the atmosphere and becoming, quite literally, a space rover. As he rockets through the sphere of dead satellites, we see the grapefruit-sized Sputnik bounce off of him, and as he flies over the Moon, he passes an Apollo landing site with the landing pad, the American flag, and a Moon rover as still as a set of toys beneath him (accompanied by a BNL billboard).

In that voyage, Wall-E becomes another sojourner, another rover wandering the universe and imbued with spirit and curiosity, persevering. His arrival on the Axiom brings him into first contact with humanity. However, those humans' lives have been reduced to a bare-bones subsistence—a Huxley-esque Brave New World of entertainment-induced unconsciousness, unaware of the passage of time, the world they abandoned 700 years before, or the very circumstances of their existence. In some ways, their lives are similar to humans in *The Matrix*, confined to their seats, staring into an artificial reality, devoid not only of purpose, but of mobility, left without a world to wander.

Wall-E and Eve, in that sense, are charged with unplugging humanity from the Matrix and returning them to reality: a bleak reality, for sure, but one with purpose and hope. To connect with that reality means reconciling with the need to be more connected to the planet we call home. As a corollary to the dangers of refusing to look up to see the asteroid coming, there is also a danger in refusing to look down, or inward, to assess the world around us, to learn how to love and cherish it, to examine the rocks, dig our hands into the dirt, and discover our home planet anew.

23 EULOGY

Life is a phenomenon that exists on the razor's edge between cold and heat. Unfortunately, man has not been content with the hazards that nature provides. He has been busy building his own funeral pyre.

—Sir Arthur Clarke, *Earthlight*

At Mr. Cooper's funeral, I read a short speech I'd written the night before. I was sixteen and the paper shook in my hands. I did my best not to make eye contact with Mrs. Cooper.

I talked about how his influence had shaped me, about how much I'd learned from that whistling mad scientist. I mentioned our hovercraft, our propeller car, the exploding cereal box we made for a short movie I created for class. I talked about how he inspired my curiosity, built up my confidence, and made me feel like I had the tools to navigate an uncertain world. After the funeral, Mrs. Cooper gave me an enormous hug.

"Guy always loved his time with you," she said. "Whenever you couldn't come over, it was like he'd missed a play date."

I felt the love in what she was saying, but immediately I felt that urge to grasp the arrow of time and turn it backwards. In truth, I had seen him less frequently by the time I went to high school. Even though we were still scheduled to see each other every Tuesday, I would often find an excuse not to come over. I was more interested in hanging out with my friends, sneaking off to the beach, being a teenager. I lived in that adolescent mindset that always assumed there was a tomorrow, a next week, a future. It had never once occurred to me that a Tuesday afternoon would come along where I couldn't go to Mr. Cooper's if I wanted to. Standing there looking at Mrs. Cooper, I wanted to go back to every afternoon I'd missed. I would've given anything for one more day with Mr. Cooper, working at his lathe or talking about his surfing robot.

This book is full of criticism, but that criticism shouldn't mask the joy, the curiosity, the thrill of discovery and the awe of what we, as a species, have accomplished. Mr. Cooper opened my understanding to that world, and even as we confront increasingly precarious global circumstances, I can't help but maintain some of the optimism he imparted to me. It's the same sort of optimism that Carl Sagan had. A sense of realism, a knowledge that humans have the intellect and capacity to solve our problems, and that the solutions to our most profound challenges are waiting in black boxes for the right question to be asked.

I don't know if there is some "madness gene" that makes us set off for the unknown. I'm more inclined to think that humans really do just have a restless curiosity. If we didn't, we wouldn't make music, sketch the wilderness, write poetry, sing songs. For some, that curiosity manifests itself as whittling away at a piece of balsa wood while sitting on a front porch rocking chair. For others, it appears as an urge to set sail. And still, for others yet, it's the desire to build a machine that can fend for itself on another world.

Civilization has always been in tension between the creative and destructive force. That's the very definition of chaos and cosmos. Disorder and order. A rover, whether operated by a human or driven by its internal programming, is always on the frontier between those two zones. It brings with it the capacity for so much beauty and so much despair. And always, it is entangled with the world from which it came.

The Perseverance rover was sent to Mars to collect samples. Its drill fills vials that it stashes internally, or leaves on the planet in various caches for a future Mars Sample Return Mission to recover. That mission is supposed to bring those samples back to Earth, delivering the first soil from Mars that wasn't delivered as the result of a catastrophic impact that hurled a rock into deep space and sent it plummeting through Earth's atmosphere in a beautiful ball of fire. Those samples, examined in the world's best labs, will reveal Mars to us like never before. But that mission has been complicated by the war in Ukraine. The dissolution of international relations—and a recent assessment that ballooned the projected cost of the mission—has forced the European Space Agency (ESA) and NASA to abandon a collaboration that would have used a Russian rocket to deliver the Mars Sample Return rover. Whatever our intentions in the stars, we have to survive long enough to fulfill them. It's possible to send rovers to other star systems, since they would be unburdened by the inconvenience of needing to stay alive—and advances in artificial intelligence may further "humanize" our robot emissaries. But by the time one of these rovers reaches the next star, who's to say that the nation that launched it, or the civilization it was a part of, or even its entire home planet, will still be there to receive the message upon return?

Every single planetary rover has carried with it an ephemeral payload of national ambitions, personal dreams, fears, interests and prejudices. India's Chandrayaan-3 lunar rover, which landed on the Moon's south pole in July of 2023, was both a monumental technical achievement and a statement of the arrival of India and the developing world in the fraternity of spacefaring (and planet-roving) nations. Just as the space race did for the US and Russia, Chandrayaan affords India a certain prestige that can be used to counter or drown more controversial narratives, such as those concerning emerging Hindu nationalism and the backsliding of the world's largest democracy into something more autocratic. The pride of success or the contributions to science should not be dismissed because of these political entanglements. The entanglements are always there, for India, China, the United States, Russia, Japan and every rover from every country and every private company to come. Every journey into space is pregnant with boundless possibility and peril, because every rover is just a product of the civilization, and the species, that built it. Our terrestrial and cosmic destinies will always be entangled.

Mr. Cooper was always someone who believed that technology was mostly amoral—subject to the ethics of those who possess it. I think I've inherited that belief from him, too. I'm not afraid of atomic energy or artificial intelligence. I'm afraid of the intentions of those who possess these tools. I'm afraid of human nature—as afraid of it as I am in awe of it. And it's for that reason that I fear what we will bring with us to other worlds.

I think I have to return, once more, to that Cosmist desire to transcend humanity; that Earth is the cradle, and perhaps we are not meant to remain the cradle forever. But we cannot leave the cradle without maturing, otherwise the benign cruelty of the universe will consume us, or the malignant cruelty of our species will devour those we encounter. That transcendence doesn't come at the expense of Earth. It arrives with our realization of the fragile beauty of our home planet, and that we are, more than anything else, Earthlings.

ACKNOWLEDGMENTS

On the day that I submitted the final draft of this book, I learned that a slow-growing cancer that began in my father's throat might have spread to his lungs—an escalation that has shifted our mindsets from "We can't wait to beat this" to "We're going to fight this, and we're holding on to hope." I was sitting at a bar when I hit "send" on the final draft of this manuscript, and then I tossed back my beer in celebration and immediately tried to fight back the tears as I contemplated the way the universe seems to like to throw dark clouds over what should be life's sweetest moments.

I wasn't going to write any acknowledgments for this book, but now I find myself wanting to take every opportunity available to express my love for my father and for everyone else who has supported, encouraged, and guided me as I've wandered through this life. I owe everything to my mother, who sought out the scientist to whom this book is dedicated and asked him to be my mentor. And I am so grateful to the Fogliadinis for inspiring this book through their work on the Lunar Roving Vehicle. I am also ever grateful to Stephanie

Gaitán for being a right-hand support for so much of my writing. More and more as I grow older, I feel empowered by the love and support of my family and friends, old and new, and I want to express my thanks to all of them. This book has been a beautiful journey of discovery and introspection.

After then-vice president Joe Biden's son Beau died from brain cancer near the end of the Obama administration, the administration introduced the Cancer Moonshot—an initiative to reduce cancer deaths, improve screening, and improve the conditions of those confronting cancer. I can't help but reflect on the idea of the "moonshot" in this acknowledgment—the idea of marshalling your forces in pursuit of a singular and audacious objective. As you read this book, you'll note the tension between reverence and terror for what technology and science have achieved, but if I have hope for my father, that hope is based on an excitement at the seismic shift in treatment that may take place as AI, nanoparticles, and other targeted treatments are applied to cancer. So many of our breakthroughs, our moonshots, have been accidents or luck, but those accidental and lucky moments only arise when we seek them. As I watch my father persevere, I hold out hope that the next breakthrough will change his life, and the lives of millions. I hold out hope for a moonshot.

BIBLIOGRAPHY

Aderin-Pocock, Maggie. *The Book of the Moon: A Guide to Our Closest Neighbor*. Abrams Image, 2019.

Al, Reinert, director. *Apollo 11: For All Mankind*. Apollo Associates, 1989.

Aleksievich, Svetlana. *Secondhand Time: The Last of the Soviets: An Oral History*. Random House, 2016.

Baker, David. *NASA Mars Rovers Manual: 1997–2013 (Sojourner, Spirit, Opportunity and Curiousity): An Insight into the Technology, History and Development of NASA's Mars Exploration Roving Vehicles*. Haynes Publishing, 2013.

Ballard, Robert. "The Titanic Discovery: Professor Robert Ballard." YouTube. *Inner Earth Institute*, 20 Apr. 2012, www.youtube.com/watch?v=5Q3eA6wYil4.

Bjornerud, Marcia. *Timefulness How Thinking like a Geologist Can Help Save the World*. Princeton University Press, 2020.

Branch, John, and Christina Goldbaum. "A Rubik's Cube, Thick Socks and Giddy Anticipation: The Last Hours of the Titan." *New York Times*, 2 July 2023, www.nytimes.com/2023/07/02/us/titan-submersible-passengers.html?name=styln-titanic-submersible®ion=TOP_BANNER&block=storyline_menu_recirc&action=click&pgtype=Article&variant=undefined.

Chris, Feltcher. "'buryings' a Man on the Moon." *NBCNews.Com*, 1 Dec. 2005, www.nbcnews.com/id/wbna3077929.

Clarke, Arthur C. *Earthlight*. ROSETTABOOKS, 2022.

"Defunctland: The History of Tomorrowland 1955." YouTube, 19 Apr. 2020, www.youtube.com/watch?v=fTGa8HIsoyg.
Dent, Steve. "NASA Goes Steampunk for Its Future Venus Probes." *Engadget*, 13 May 2021, www.engadget.com/2017-08-28-nasa-futuristic-probes-niac.html.
Dunbar, Brian. "Konstantin E. Tsiolkovsky." *NASA*, 5 June 2013, www.nasa.gov/audience/foreducators/rocketry/home/konstantin-tsiolkovsky.html.
Frostensen, Sarah. "Episode 2: First Steps: Sojourner." *NASA*, www.jpl.nasa.gov/podcasts/on-a-mission-season-1/season-4-mars-rovers/episode-2-first-steps-sojourner. Accessed 24 July 2023.
"Girl Who Named Mars Rover Stays down to Earth." *New York Times*, 14 July 1997, www.nytimes.com/1997/07/14/nyregion/girl-who-named-mars-rover-stays-down-to-earth.html.
Gorman, Alice. *Dr Space Junk vs the Universe: Archaeology and the Future*. MIT Press, 2020.
Hayden, Tyler. "Santa Barbara Racers Chase Land Speed Records." *Santa Barbara Independent*, 23 Aug. 2018, www.independent.com/2018/08/23/santa-barbara-racers-chase-land-speed-records/.
"He Always Dreamed of Stepping Foot on the Moon. He Made It There after His Death." *CBS58*, www.cbs58.com/news/he-always-dreamed-of-stepping-foot-on-the-moon-he-made-it-there-after-his-death. Accessed 24 July 2023.
Hevesi, Dennis. "Betty Skelton, Air and Land Daredevil, Dies at 85." *New York Times*, 11 Sept. 2011, www.nytimes.com/2011/09/11/us/11skelton.html.
Johnson, Sarah Steward. *Sirens of Mars: Searching for Life on Another World*. Penguin Books, 2021.
Kolbert, Elizabeth. *Sixth Extinction: An Unnatural History*. Holt Paperbacks, 2024.
Leasca, Stacey. "NASA's Perseverance Rover Is Tweeting from Mars and It's Both Hilarious and Educational." *Travel + Leisure*,

23 Feb. 2021, www.travelandleisure.com/trip-ideas/space-astronomy/perseverance-rover-first-color-image-mars.

Ludolph, Emily. "Ed Dwight Was Set to Be the First Black Astronaut. Here's Why That Never Happened." *New York Times*, 16 July 2019, www.nytimes.com/2019/07/16/us/ed-dwight-was-set-to-be-the-first-black-astronaut-heres-why-that-never-happened.html.

Magazine, Smithsonian. "My Other Ride Is a Spaceship." *Smithsonian.Com*, 1 June 2019, www.smithsonianmag.com/air-space-magazine/my-other-ride-is-spaceship-astronauts-corvettes-180972234/.

Morris, Austin. "Tyranny of the Rocket Equation - KMI." *Kall Morris Incorporated*, 27 Feb. 2023, www.kallmorris.com/columns/tyranny-of-the-rocket-equation#:~:text=And%20herein%20lies%20the%20tyranny,for%20this%20increase%20in%20size.

"The Origins of Polish Sci-Fi & Jerzy Żuławski's Legacy: Feature." *Culture.Pl*, culture.pl/en/feature/the-origins-of-polish-sci-fi-the-legacy-of-jerzy-zulawski. Accessed 24 July 2023.

Rees, Martin J., and Donald Goldsmith. *The End of Astronauts: Why Robots Are the Future of Exploration*. Harvard University Press, 2022.

Ridley, Scott, director. *The Martian*. Twentieth Century Fox Home Entertainment, 2016.

Riley, Christopher, et al. *Lunar Rover Manual: 1971–1972 (Apollo 15–17 ; LRV1-3 & 1G Trainer): Owner's Workshop Manual*. Haynes, 2012.

Robinson, Kim Stanley. *The Ministry for the Future*. Orbit, 2022.

Ryan, White, director. *Good Night, Oppy*. Amazon Studios, 2023.

Seams, Clayton, "Here's Why so Many Astronauts Have Owned Chevrolet Corvettes," *Driving*, driving.ca/auto-news/entertainment/heres-why-so-many-astronauts-have-owned-chevrolet-corvettes. Accessed 24 July 2023.

Sinclair, Stewart L. "Fahrenheit 2017." Guernica, 1 May 2018, www.guernicamag.com/fahrenheit-2017/.

Stanton, Andrew, et al. *Wall-E*. Pixar Animation Studios. 2008.

Stevis-gridneff, Matina, and Karam Shoumali. "Everyone Knew the Migrant Ship Was Doomed. No One Helped." *New York Times*, 1 July 2023, www.nytimes.com/2023/07/01/world/europe/greece-migrant-ship.html.

Swift, Earl. *Across the Airless Wilds: The Lunar Rover and the Triumph of the Final Moon Landings*. Mariner Books, 2022.

Afanassieff, Jean. *Tank on the Moon*. Corona Films. 2007.

"Timeline: How the Migrant Boat Tragedy Unfolded off Greece." *Al Jazeera*, 17 June 2023, www.aljazeera.com/news/2023/6/16/timeline-how-the-refugee-boat-tragedy-unfolded-off-greece.

Traub, Alex. "Carolyn Shoemaker, Hunter of Comets and Asteroids, Dies at 92." *New York Times*, 1 Sept. 2021, www.nytimes.com/2021/09/01/science/space/carolyn-shoemaker-dead.html.

Tyson, Neil deGrasse, and Avis Lang. *Accessory to War: The Unspoken Alliance between Astrophysics and the Military*. W.W. Norton & Company, 2018.

Young, Anthony. *Lunar and Planetary Rovers the Wheels of Apollo and the Quest for Mars*. Springer/Praxis, 2007.

Heron, Gil Scot. "Whitey on the Moon," YouTube, 20 July 2019, https://www.youtube.com/watch?v=3nzoPopQ7V0. Accessed 24 July 2023.

Żuławski, Jerzy. *The Lunar Trilogy*. Zmok Books, 2020.

INDEX